Experimental
and
Analytical Bakery

THE AUTHOR

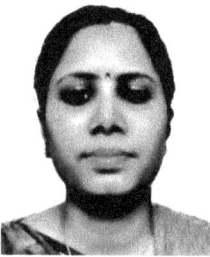

Dipti Sharma is an Assistant Professor in the Department of Food Technology at a College of University of Delhi. She has her specialization in Food Science and Technology & have completed B.A. Sc. & M.Sc. in Food Technology from University of Delhi and G.J.U of Science and Technology respectively. She is pursuing Ph.D in Food Technology from Dr. K.N. Modi University, Rajasthan. She has completed a Certificate Course in Consumer Protection and a PG Diploma in Food Safety and Quality Management from IGNOU. She has more than 8 years of industry and teaching experience.

She is a life member of the Association of Food Scientist and Technologists of India, Society of Indian Bakers and ISTE. Besides coordinating a number of workshops and seminars, she has presented papers at the national and international conferences. She is also in the reviewer/editorial board of journal.

Experimental and Analytical Bakery

Dipti Sharma

2016

Daya Publishing House®

A Division of

Astral International (P) Ltd

New Delhi 110 002

Cataloging in Publication Data—DK
Courtesy: D.K. Agencies (P) Ltd.
<docinfo@dkagencies.com>

Sharma, Dipti (*Assistant professor of food technology*), **author.**

Experimental and analytical bakery / Dipti Sharma.
pages ; cm

ISBN 9789351306740 (International Edition)

1. Baked products. 2. Baking. 3. Confectionery.
4. Food—Analysis.   I. Title.

DDC 664.752   23

Published by   : **Daya Publishing House**®
                             A Division of
                             Astral International Pvt. Ltd.
                             – ISO 9001:2008 Certified Company –
                             4760-61/23, Ansari Road, Darya Ganj
                             New Delhi-110 002
                             Ph. 011-43549197, 23278134
                             E-mail: info@astralint.com
                             Website: www.astralint.com

Laser Typesetting   : **Classic Computer Services**, Delhi - 110 035

Printed at   : **Replika Press Pvt. Ltd.**

The book is dedicated to my guruji "Shri Ashutosh Maharajji" and son, my mother for inspiration; to my father and brothers for their supports; to my loving husband for constant encouragement and motivation.

Acknowledgement

I am thankful to my mentor Prof. Rita Bakshi and my students for encouraging me to write a book with both product recipe's and analytical procedure.

I am thankful to Astral International (P) Ltd. for publishing this book.

Dipti Sharma

Foreword

I am very happy to learn that a book titled *"Experimental and Analytical Bakery"* written by Dipti Sharma, Assistant Professor, University of Delhi is being published. In this book the author has incorporated various recipes of bakery products and the analytical procedure to the conduct the tests.

This well written, concise manual apart from being an excellent academic tool, it shall also find extensive application in the food industry as a guide for chemical analysis aimed at regulation compliance and evaluation of nutritional values of foods.

The information is given in simple and easy language and the well defined information given in the book will not only be a useful for students, researchers and individuals not only in the field of food and nutrition but also for food technology and other allied fields.

I, congratulate everyone associated with the publication of this book and extend best wishes for successful publication of the book.

Prof. (Dr.) Rita Bakshi
Former Principal, Ginni Devi Modi
Girls, P.G. College, Modinagar
Meerut

Preface

The need for chemical analysis of food has increased in recent years. Growing emphasis in the industry is given on nutritional value of foods, value addition and on labeling regulations pertaining to display nutritional values of food product on the package, the latter requires analysis of dietary components. This requires not only the correct recipe of making the product but also accurate analysis of food for its constituents.

This book titled *"Experimental and Analytical Bakery - Manual"* is divided into two parts: Part I covers the recipes of bakery products and Part II covers the analytical procedures for performing the food analysis experiments. Prior knowledge and clear concepts about an experiment help the students to perform the practical properly, efficiently and safely. The aim of this book is to give students the experience in performing food analysis experiments, analyzing data and reporting their findings. So, an attempt has been made to impart information about recording experiments in a proper format and correct representation of data for the students in this single volume.

Feedback and suggestions from colleagues as well as students to make the book more useful are most welcome.

Dipti Sharma
Assistant Professor
University of Delhi

Contents

Acknowledgement   vii

Foreword   ix

Preface   xi

PART-I : EXPERIMENTAL BAKERY AND CONFECTIONERY

EXPERIMENTAL BAKERY

1. Preparation of Bread   3
2. Preparation of Bun   5
3. Preparation of Milk Bread   9
4. Preparation of Bread By Sponge and Dough Method   11
5. Preparation of Whole Wheat Flour Bread   13
6. Preparation of Brown Bread   15
7. Preparation of Danish Pastry   17
8. Preparation of Potato Barn Bread   19
9. Preparation of Dough Nut   21
10. Preparation of Jeera Toast   22
11. Preparation of Tutti Fruity Toast   24
12. Preparation of Mava Toast   25
13. Preparation of Pizza   26
14. Preparation of Pitza   28

CONFECTIONERY

1. Preparation of Nankhatai   30
2. Preparation of Vanilla Biscuit   32
3. Preparation of Chocolate Biscuit   34
4. Preparation of Sweet and Salty Biscuits   36
5. Preparation of Melting Moments Cookies   37
6. Preparation of Cherry Short or Tutti Fruity Biscuit   39
7. Preparation of Coconut Cookies   41
8. Preparation of Falhari Nankhatai (for Fasting)   44

9. Preparation of orange Biscuit — 45
10. Preparation of Square Nut Biscuit — 47
11. Preparation of Pineapple Cookie — 49
12. Preparation of Peanut Cookie — 51
13. Preparation of Cherry Cookie — 52
14. Preparation of Jam Sandwiched Buns — 54
15. Preparation of Nut Ring Biscuit — 56
16. Preparation of Falhari Biscuit — 58
17. Preparation of Coconut Castle — 59
18. Preparation of Cinnamon Crescent — 61
19. Preparation of Tricolour Biscuits — 63
20. Preparation of Masala Biscuit — 65
21. Preparation of Atta (Whole Wheat Flour) Biscuit — 66
22. Preparation of Sweet Biscuit — 67
23. Preparation of Ester Biscuit — 68
24. Preparation of Salty and Spicy Cashew Shaped Biscuit — 69
25. Preparation of Vegetable Puff — 70
26. Preparation of Medallin Cake — 72
27. Preparation of Queen Cake — 74
28. Preparation of Cherry Cake — 76
29. Preparation of Colour Pastry — 78
30. Preparation of Pineapple and Cherry Upside Down Cake — 79
31. Preparation of Egg Less Cup Cake/Cake — 81
32. Preparation of Date and Walnut Cake — 83
33. Preparation of Coconut Macroons — 84
34. Preparation of Brownie Cake — 85
35. Preparation of Cherry Knob — 86
36. Preparation of Dry Fruit Cake — 87
37. Preparation of Welse Cheese Cake — 89
38. Preparation of Date and Ginger Cake — 90
39. Preparation of Cream Rolls — 91
40. Preparation of Coffee Buns — 92
41. Preparation of Chocolate — 93
42. Preparation of Sponge Cake with Icing — 94
43. Preparation of Chocolate Fudge — 96
44. Preparation of Icing and Fondant — 97
45. Preparation of Plain Butter Sponge Cake — 100
46. Preparation of Chocolate Cake — 102
47. Preparation of Sponge Cake with Pineapple Icing — 103

48. Preparation of Chocolate Sponge Cake with Icing   105
49. Preparation of Tutti Fruity Cake   107
50. Preparation of Marble Cake   109

PART-II: ANALYTICAL BAKERY

EXPERIMENTS

1. Determination of moisture content in food products by hot air oven- drying method   113
2. Determination of ash content in flour sample   115
3. Determination of acid insoluble ash (AIA) content in flour sample   118
4. Determination of protein content in food products by kjeldahl method   119
5. Determination of falling number of wheat flour sample.   123
6. Determination of thousand kernel weight of wheat grain sample   125
7. Determination of gluten content in the wheat flour sample.   126
8. Estimation of volume and specific gravity of flour and cornflour   128
9. Qualitative estimation of flour refineness by peckar colour test   129
10. Determination of sedimentation value of the flour   130
11. Determination of yeast quality by its dough rising capacity   131
12. Determination of starch in cereal grains by acid hydrolysis method   132
13. Determination of salt content in finished bakery products   135
14. Determination of specific gravity of oils and fats   137
15. Determination of saponification value of oil sample   139
16. Determination of iodine value of oil and fat sample (WIJS METHOD)   141
17. Determination of salt content in butter   143
18. Estimation of peroxide value of oil sample   144
19. Determination of crude fat in foods by soxhlet extraction method   145
20. Determination of pH of food products by using pH meter   147
21. Determination of pH of the aqueous extract of the sample   149
22. Determination of purity of sodium chloride (NaCl) Salt   150
23. Determination of refractive index of fat & oil sample   151
24. Determination of acid value in the fat or oil sample   154
25. Determination of free fatty acids in the fat or oil sample   156

Part-I
Experimental Bakery
and
Confectionery

EXPERIMENTAL BAKERY
Practical No. 1

Aim: Preparation of Bread.

Procedure

1. Sieve the flour.
2. Take about 1/10th of total requirement of luke warm water (37°C/ 98.6°F) water. Add a part (about 1/5th of total requirement) of sugar in it.
3. Add the crumbled yeast and allow to rest aside (for about 5 to 10 min) till it disintegrates and starts to float on the water. If yeast does not disintegrate, it is understood that the yeast is of poor quality. Hence, throw it away and follow the same procedure with fresh yeast.
4. Add sufficient flour to make thin paste and whisk it to incorporate some air, leave it for 10 to 15 min, during which the paste will arise, that helps in vigorous yeast action.
5. Dissolve salt and left over sugar in remaining water and strain to remove extraneous matter. If formula contains milk powder, add it at this stage in such a way that it does not make lumps.
6. Add this water into flour, roughly mix, add the yeast paste and knead well to prepare a smooth dough. Add shortening at the last stage of mixing and make clear dough.
7. Cover the dough with wet cloth and keep it aside for a stipulated time at 26.6°C (80°F) and 75 per cent RH for bulk fermentation.
8. Press out the gas produced after 2/3rd of bulk fermentation time; which is known as knock back.
9. Divide the dough into the desired size pieces (the weight of piece must be 12 per cent extra than that desirable bread weight) that is termed as dividing/scaling.
10. Round each dough piece and allow it to rest for 10 min, technically it is termed as rounding and intermediate proofing. Use sharp cutter for diving, do not stretch with hand.
11. Mould the dough pieces, place it in a greased bread tin, cover the lid and then rest for about 1 hr for final proofing at 35 to 36°C and 80 per cent RH.
12. Bake it at 230°C (450°F) for about 25-30 minutes.
13. Remove the baked bread from the tin or case on mesh tray and apply a thin layer of melted shortening on surface of the bread.
14. Allow it to cool for about 1 hour and then pack it in polyethylene bag and store it at room temperature.

Observations

S.No.	Ingredients	Weight of Ingredients	%	Cost/kg (Rs.)	Cost (Rs.)
1.	Flour	250 g	100		
2.	Sugar	12.5g	5		
3.	Yeast	5g	1.2		
4.	Salt	5g	1.5-2		
5.	H.V.O	5g	2		
6.	Water	150ml	60		
Total		**427.50**			

Weight of dough   =   427.5 g

Cost of dough   =   …… Rs.

Yield   =   …… gm

Note your sensory observations below:

External	Internal
• Crust Colour	• Crumb Colour
• Volume Gain	• Symmetry of Form
• Aroma	• Sensory

Calculations

1. Over Head Cost (O.H.C.)	1. Baking-Cooling losses
	= Wt. of dough – Wt. of bread
2. Total cost = Ingredient cost + O.H.C.	2. % (Baking-Cooling) loss

Practical No. 2

Aim: Preparation of Bun.

Procedure

1. Sieve the flour
2. Take about 1/10th of the total required luke warm water and then add about 1/5th of the sugar in it.
3. Then add the crumbled yeast to it and allow to rest for about 5-10 minutes till it floats.
4. Add sufficient flour to make thin paste and whisk it to incorporate some air and leave it for about 10-15 minutes.
5. Dissolve alt and left over sugar in remaining water and strain it to remove extraneous matter.
6. Add this water into flour, roughly mix it and add the yeast paste and knead well to prepare a smooth dough. Add shortening at the last stage of mixing and make a clear dough. Cover the dough and keep it for bulk fermentation at 26.6 °C and 75 R.H for about 1 to 1½ hour.
7. Press out the gas produced after bulk fermentation known as knock-back.
8. Now shape it in about 50 gm pieces and round it
9. Place all there in a greased tray land proof till the desired volume achieved. Now flatten them with flat hand.

Observations

S.No.	Ingredients	Weight of Ingredients	%	Cost/kg (Rs.)	Cost (Rs.)
1.	Flour	100	250		
2.	H.V.O	2	5		
3.	Sugar	5	12.5		
4.	Yeast	1 to 2	5		
5.	Salt	1 to 2	3.75		
6.	Water	60	150ml		
	Total		426.25		

Weight of dough = 426.25 gm
Cost of dough   = ... Rs.
Yield   = ... gm

Note your sensory observations below:

External	Internal
• Crust Colour	• Crumb Colour
• Volume Gain	• Symmetry of Form
• Aroma	Sensory

Calculations

1. Over Head Cost	1. Baking-Cooling losses = Wt. of dough – Wt. of bread
2. Total cost = Ingredient cost + O.H.C.	2. % (Baking-Cooling) loss

Alternate Recipe-2

Aim: Preparation of Bread and Bun.

Procedure

1. Mix Yeast with 1 spoon of sugar and 3-4 spoons of maida and make a paste by mixing it thoroughly with water.

2. Put this yeast suspension at 37°C and 70 per cent RH for 15min, mix salt, SMP and Sugar solutions with major proportion of water, then add improver to the remaining maida and sieve it thrice.

Preparation of Dough

3. In the planetary mixer bowl, add table salt, SMP and sugar solution and add maida and yeast suspension to it and mix it at speed of 1.

4. After the dough start getting form then increase the speed to 2 and followed by speed to 3.

5. Add vegetable oil and then again mix it at speed of 2 and 3 till the dough ball is formed.

6. Grease the bowl and put dough in it and keep it in a proover for 15 mins at 37°C and 70 per cent RH.

7. Roughly divide the dough in parts- 480g for bread and 80 g for buns. Keep it for rest for 15 mins.

8. Mould into final shape and place bread loafs into moulds and keep it for final proofing at 42°C for 45 minutes and give a coating of milk and/or oil.

9. Take out the bread loaf from the proover and bake it at 250°C for 45 minutes. Then make balls of buns and put it in a greased (tin) tray and bake it at 180°C for 15-20 minutes.

10. Cool the breads/buns after baking and then slice it and pack it.

11. Write the observations and calculations.

Note

- The glaze on bread is due to gelatinization of starch when humidity is high.
- Milk and oil is applied on the surface of dough before baking.
- After baking oil is applied on the surface of bread loaf add bins to give a glaze on it.

Formulation

S.No.	Ingredients	Weight of Ingredients (gm)	%	Cost/kg (Rs.)	Cost
1.	Flour	250 g			
2.	Sugar	5 g			
3.	Yeast	5g			
4.	Salt	3.75g			
5.	Fat/H.V.O	5g			
6.	Improver	2.5g			
7.	SMP	3.75g			
8.	Water	150ml			

Weight of dough   = ... g

Cost of dough   = ... Rs.

Yield   = ... g

Note your sensory observations below:

External	Internal
• Crust Colour	• Crumb Colour
• Volume Gain	• Symmetry of Form
• Aroma	• Sensory

Calculations

1. Over Head Cost (O.H.C.)	1. Baking-Cooling losses
	= Wt. of dough – Wt. of bread
2. Total cost = Ingredient cost + O.H.C.	2. % (Baking-Cooling) loss

Practical No. 3

Aim: Preparation of Milk Bread.

Procedure

1. Sieve the flour.
2. Take about $1/10^{th}$ of the total required luke warm water and then add a about 1/5the of the sugar in it.
3. Then add the crumbled yeast to it and allow to rest for about 5-10 minutes till it floats.
4. Add sufficient flour to make thin paste and whisk it to incorporate some air and leave it for about 10- 15 minutes. Dissolve salt and left over sugar in remaining water and strain to remove extraneous matter. If formula contains milk powder, add it at this stage in such a way that it does not make lumps. Add this water into flour, roughly mix add the yeast-paste and knead well to prepare a smooth dough. Add shortening at the cast stage of mixing and make a clear dough.
5. Cover the dough and keep it aside for about 1 to 1½ hours at 26.6°C and 75 R.H. for bulk fermentation.
6. Press out of gas produced after bulk fermentation known as knock-back.
7. Divide the dough into the desired size and then round them and allow it to rest for about 10 minutes.
8. Mould the dough pieces and place them in a greased tin, cover the lid and rest for about 1 hour for proofing at 35 to 36 °C and 80 R.H. to 85 R.H.
9. Bake it at 205°C for about 20 to 30 minutes.
10. Allow the bread to cool and then pack it into polyethylene bags and store it at room temperature.

Observations

S.No.	Ingredients	Weight of Ingredients (gm)	%	Cost/kg (Rs.)	Cost
1.	Flour	250	100		
2.	H.V.O.	20	8		
3.	Milk Powder	15	6		
4.	Sugar	35	15		
5.	Yeast	5	2		
6.	Water	150 ml	60-65		
7.	Salt	3.5	2		
	Total				

Weight of dough   =   508.50 gm

Cost of dough   =   ………. Rs.

Yield   =   ……….. g

Note your sensory observations below:

External	Internal
• Crust Colour	• Crumb Colour
• Volume Gain	• Symmetry of Form
• Aroma	• Sensory

Calculations

1. Over Head Cost (O.H.C.)	1. Baking-Cooling losses
	= Wt. of dough – Wt. of bread
2. Total cost = Ingredient cost + O.H.C.	2. % (Baking-Cooling) loss

Practical No. 4

Aim: Preparation of Bread by Sponge and Dough Method.

Procedure

1. Sieve the flour twice.
2. Take about 60 per cent of flour, water and sugar and 100 per cent yeast for preparation of sponge.
3. In a small bowl, take luke warm water and then add yeast and sugar and little flour to prepare a flying ferment.
4. After getting a flying ferment add it to the flour (60 per cent) and make it into a smooth and loose sponge and then allow it to proof for about 30 – 45 minutes.
5. Now mix the remaining raw materials to the dough along with water and salt.
6. Knead it quickly to smooth dough.
7. Smooth the shortening and then mix it with the dough and then allow it to ferment.
8. After fermentation cut the dough into prescribed weight and then round it followed by intermediate proofing.
9. Now mould the dough pieces and place them in well greased pairs followed by final proofing.
10. Bake it at 230°C for about 25-30 minutes.
11. Depan the bread and grease the surface for a good finish only on the surface.
12. Allow the bread to cool, followed by cutting and packing in polyethylene bags.

Observations

S.No.	Ingredients	Weight of Ingredients (gm)		%	Cost/kg (Rs.)	Cost (Rs.)
		Sponge	Dough			
1.	Flour	300	200	100		
2.	Sugar	15	10	5		
3.	Yeast	5	–	1 – 2		
4.	Salt	–	7.5	1.5 – 2		
5.	Shortening	–	10	2		
6.	Water	18 – 20	120	60		
	Total	520	347.5			
		867.5gm				

Wt. of dough   =   867.5 gm
Cost of dough   = ... Rs.
Yield   = ... g

Note your sensory observations below:

External	Internal
• Crust Colour	• Crumb Colour
• Volume Gain	• Symmetry of Form
• Aroma	Sensory

Calculations

1. Over Head Cost =	1. Baking-Cooling losses = Wt. of dough–Wt. of final product
2. Total cost = Ingredient cost + O.H.C.	2. % (Baking-Cooling) loss

Practical No. 5

Aim: Preparation of Whole Wheat Flour Bread.

Procedure

1. Sieve the wheat flour (whole meal flours).
2. Take about $1/10^{th}$ of total requirement of luke warm water (37°C/ 98.6°F) water. Add a part (about $1/5^{th}$ of total requirement) of sugar in it.
3. Add the crumbled yeast and allow to rest aside (for about 5 to 10 minutes.) till it disintegrates and starts to float on the water. If yeast does not disintegrate, it is understood that the yeast is of poor quality. Hence, throw it away and follow the same procedure with fresh yeast.
4. Add sufficient flour to make thin paste and whisk it to incorporate some air, leave it for 10 to 15 min, during which the paste will arise, that helps in vigorous yeast action.
5. Dissolve salt and left over sugar in remaining water and strain to remove extraneous matter. If formula contains milk powder add it at this stage in such a way that it does not make lumps.
6. Add this water into wheat flour, roughly, mix. add the yeast paste and knead well to prepare a smooth dough Add shortening at the last stage of mixing and make a clear dough.
7. Cover the dough with wet cloth and keep it aside for a stipulated time at 26.6°C and 75 per cent for bulk fermentation.
8. Press out the gas produced after $2/3^{rd}$ of bulk fermentation time, which is known as knock back.
9. Divide the dough into the desired size pieces (the weight of piece must be 12 per cent extra than that desirable bread weight) that is termed as dividing /scaling
10. Round each dough piece and allow to rest for 10 minutes, technically it is termed as rounding and intermediate proofing. Use sharp cutter for dividing, do not stretch with hand.
11. Mould the dough pieces and place it in a greased bread tin. Cover the lid and then rest for about 1 hour for final proofing at 35 to 36°C and 80 per cent R.H.
12. Bake it at 230°C (450°F) for about 25-30 minutes.
13. Remove the baked bread from the tin on sieve and apply a light out of melted shortening on surface of the bread.
14. Allow it to cool for about 1 hour and then pack it in polyethylene bag and store it at room temperature.

Observations

S.No.	Ingredients	Weight of Ingredients (gm)	%	Cost/kg (Rs.)	Cost (Rs.)
1.	Wheat Flour	125	100		
2.	Ghee	2-5	2		
3.	Sugar	6-25	5		
4.	Salt	2-5	2		
5.	Yeast	2-5	2		
6.	Water	73	60		
7.	M. Powder	2-5	2		
	Total	**266.25**			

Weight of dough   = ... g
Cost of dough   = ... Rs.
Yield   = ... g

Note your sensory observations below:

External	Internal
• Crust Colour	• Crumb Colour
• Volume Gain	• Symmetry of Form
• Aroma	Sensory

Calculations

1. Over Head Cost =	1.Baking-Cooling losses
	= Wt. of dough–Wt. of final product
2. Total cost = Ingredient cost + O.H.C.	2. % (Baking-Cooling) loss

Practical No. 6

Aim: Preparation of Brown Bread.

Procedure

1. Sieve 50 per cent wheat flour and 50 per cent flour.
2. Take about 1/10th of total requirement of luke warm water (37°C). Add a part (about 1/5th of total requirement) of sugar in it.
3. Add the crumbled yeast and allow to rest aside (for about 5 to 10 minutes.) till it disintegrates and starts to floats and if yeast does not disintegrate, it is understood that the yeast is of poor quality. Hence, throw it away and follow the same procedure with fresh yeast.
4. Add sufficient flour to make thin paste and whisk it to incorporate some air. Leave it for 10-15 minutes. during which the paste will arise, that helps in vigorous yeast action.
5. Dissolve salt and left over sugar in remaining water and strain to remove extraneous matter. If formula contains milk powder; add it at this stage in such a way that it does not make lumps.
6. Add this water into flour, roughly mix add the yeast paste and knead well to prepare a smooth dough. Add shortening at the last stage of mixing and make a clean dough.
7. Cover the dough with wet cloth and keep it aside for a stipulated time at 26.6°C (80°F) and 75 per cent R.H. for bulk fermentation.
8. Press out the gas produced after 2/3rd or bulk fermentation time, which is known as knock back.
9. Divide the dough into the desired size pieces (the weight of piece must be 12 per cent extra than that desirable bread weight) that is termed as dividing/scaling.
10. Round each dough piece and allow to rest for 10 minutes, technically, it is termed as rounding and intermediate proofing. Use sharp cutter for dividing, do not stretch with hand.
11. Mould the dough pieces, place it in greased bread tin, cover the lid and then rest for about 1 hour for final proofing at 35 to 36□C and 80 per cent R.H.
12. Bake it at 230°C for about 25-30 minutes.
13. Remove the baked bread from the tin and sieve and apply a light coat of melted shortening on surface of the bread.
14. Allow it to cool for about 1 hour and then pack it in polyethylene bag and store it at room temperature.

Observations

S.No.	Ingredients	Weight of Ingredients (gm)	%	Cost/kg (Rs.)	Cost (Rs.)
1.	Wheat Flour	125	50		
2.	Shortening	5	2		
3.	Flour	125	50		
4.	Sugar	12.5	5		
5.	Salt	5	2		
6.	Yeast	4	1.5		
7.	M.powder	5	3-5		
8.	Water	200 ml	133		

Weight of dough   = ... g
Cost of dough   = ... Rs.
Yield   = ... g

Note your sensory observations below:

External	Internal
• Crust Colour	• Crumb Colour
• Volume Gain	• Symmetry of Form
• Aroma	Sensory

Calculations

1. Over Head Cost =	1.Baking-Cooling losses = Wt. of dough–Wt. of final product
2. Total cost = Ingredient cost + O.H.C.	2. % (Baking-Cooling) loss

Practical No. 7

Aim: Preparation of Danish Pastry.

Procedure

1. Prepare dough similar to bread. The egg is added along with salt and beaten nicely. Mix the dough with minimum kneading.
2. Allow it to ferment for 15 to 50 minutes.
3. Fold the dough with fat layering as similar to mathri dough.
4. Rest the dough in cool place for about 30 minutes.
5. Get ready the filling. And chop fruits nuts finely.
6. Roll the dough into a rectangle to a desired thickness.
7. Spread the filling leaving apart about 0.5cm space uncovered towards end. Sprinkle dry fruit mixture followed by cardamom nutmeg mixture.
8. Brush the ends with egg wash.
9. Roll the dough into stick form. Be careful that, the ends get stick properly. If reg. chill the dough.
10. Cut the roll to about 1.25 cm thickness with the help of thread.
11. Arrange in a tray and allow to ferment until double in size and bake.
12. Brush with sugar syrup immediately upon drawing from the oven and place on cooling rack.

Observations

S.No.	Ingredients	Weight of Ingredients (gm)	%	Cost/kg (Rs.)	Cost (Rs.)
1.	Flour	250	100		
2.	Shortening	25	10		
3.	Folding Ghee	60	25		
4.	Yeast	7.5	3		
5.	Salt	3.75	1.5		
6.	Milk	35	15		
7.	Milk Powder	6.25	2.5		
8.	Water	100	40		
9.	Essence	2.5	1		
10.	Sugar	37.5	1.5		
11.	Kaju	25	10		
12.	Grapes	25	10		
13.	Tutty fruity	50	20		
14.	Icing sugar	75	30		
15.	Ghee	50	20		
16.	Cardamom/Nutmeg	2.5	1		
17.	Sugar (Whole)	37.5	15		
18.	Water	25	10		
	Total	817			

Weight of dough   =   867.5 gm
Cost of dough   = ... Rs.
Yield   = ... g

Note your sensory observations below:

External	Internal
• Crust Colour	• Crumb Colour
• Volume Gain	• Symmetry of Form
• Aroma	• Sensory

Calculations

1. Over Head Cost =	1. Baking-Cooling losses = Wt. of dough–Wt. of final product
2. Total cost = Ingredient cost + O.H.C.	2. Baking-Cooling loss (%)

Practical No. 8

Aim: Preparation of Potato Barn Bread.

Procedure

Part-1 (Preparations before 24 hours of dough preparation).

1. Boil the potato, peel off and mesh.
2. Add yeast and sugar into water. Rest for 10 minutes.
3. Add it into flour.
4. Mix the meshed potato and rest for 20 to 24 hour.

Part-2

5. Add salt and sugar into water.
6. Add this sugar-salt solution into flour and mix roughly.
7. Add the slurry of part-1 and prepare dough.
8. Mix the shortening and knead thoroughly to obtain smooth dough.
9. Now follow the bread making process as shown. However, it requires about 2 hour proofing period.

Observations

S.No.	Ingredients	Weight of Ingredients (gm)	%	Cost/kg (Rs.)	Cost (Rs.)
(Part -1)					
1.	Steamed Potato	300	3		
2.	Flour	700	7		
3.	Yeast	50	415		
4.	Sugar	10	2		
5.	Water	1000	–		
(Part – 2)					
6.	Flour	9300	93		
7.	Sugar	480	4.8		
8.	Salt	200	2		
9.	Shortening	200	2		
10.	Yeast	50	If necessary		
11.	Water	5400	If necessary		

Weight of dough   = ... g   Yield of Bread = ... g
Cost of dough   = ... Rs.   Cost of Bread = ...Rs./ 180g

Note your sensory observations below:

External	Internal
• Crust Colour	• Crumb Colour
• Volume Gain	• Symmetry of form
• Aroma	• Sensory

Calculations

1. Over Head Cost =	1. Baking-Cooling losses = Wt. of dough–Wt. of final product
2. Total cost = Ingredient cost + O.H.C.	2. Baking-Cooling loss (%)

Practical No. 9

Aim: Preparation of Dough Nut.

Procedure

1. Add nutmeg powder, cardamom powder and milk powder into flour and sieve twice.
2. Prepare a dough similar to bread preparation.
3. Allow to rest (about 40 minutes).
4. Sheet the dough to desired thickness (0.5 to 1 cm) and cut with doughnut cutter.
5. Proof for about 30 minutes.
6. Fry in a boiled oil.
7. Allow to cool till becomes slightly warm, roll in a sugar.

Observations

S.No.	Ingredients	Weight of Ingredients (gm)	%	Cost/kg (Rs.)	Cost (Rs.)
1.	Flour	100	100		
2.	Shortening	15	15		
3.	Sugar	15	15		
4.	Yeast	2.5	2.5		
5.	Salt	1	1		
6.	Milk Powder	5	5		
7.	Water	60	60		
8.	Cardamom	2.5	2.5		
9.	Nutmeg				

Note your sensory observations below:

External	Internal
• Crust Colour	• Crumb Colour
• Volume Gain	• Symmetry of form
• Aroma	• Sensory

Calculations

1. Over Head Cost =	1. Baking-Cooling losses = Wt. of dough–Wt. of final product
2. Total cost = Ingredient cost + O.H.C.	2. Baking-Cooling loss (%)

Practical No. 10

Aim: Preparation of Jeera Toast.

Procedure

1. Sieve the flour twice
2. Add milk powder to flour and mix it properly.
3. In $1/10^{th}$ of the total requirement of luke warm water add $1/5^{th}$ of the sugar followed by addition of yeast and small quantity of flour.
4. Dissolve the left over sugar and salt in the remaining water and dissolve it completely.
5. Now add this solution after filtering into flour, with continuous mixing and then add the paste of yeast and at last mix the Jeera seeds into the dough.
6. After getting a smooth dough, add shortening into it after properly making it into a soft- paste.
7. Divide the dough into 400 gm pieces after a proofing of about 45 to 60 min at 86.6°C and 75 per cent relative humidity.
8. Place the pieces into greased trays and bake it into an oven at 205°C for about 20-25 minutes.
9. The baked toast loaf is then allowed to cool until it acquire room temperature. Arrange each slice on baking sheet separately and dry in oven at 80 to 90°C for about 1 to 1¼ hour.
10. Cool the product and then packaging is done in polyethylene bags followed by storage at room temperature.

Observations

S.No.	Ingredients	Weight of Ingredients (gm)	%	Cost/kg (Rs.)	Cost (Rs.)
1.	Flour	250	100		
2.	Shortening	15	6-8		
3.	Sugar	25	10-12		
4.	Yeast	4 (Wet), 2 (Dry)	1.5 -2.5		
5.	Water	150	60-65		
6.	Cumin / Jeera	5	2-3		
7.	Salt	4	1-3		
8.	Milk Powder	7.5	2-6		
9.	Liquid Glucose	5	1-4		
	Total	**463.5 gm**			

| Weight of dough = 463.5 gm | Wt. after first backing = ...g |
| Cost of dough = ...Rs. | Yield = ...g |

Note your sensory observations below:

External	Internal
• Crust Colour	• Crumb Colour
• Volume Gain	• Symmetry of form
• Aroma	• Sensory

Calculations

1. Over Head Cost =	1. Baking-Cooling losses
	= Wt. of dough–Wt. of final product
2. Total cost = Ingredient cost + O.H.C.	2. % Baking-Cooling loss

Practical No. 11

Aim: Preparation of Tutti Fruity Toast.

Procedure

1. Mix the powdered ingredients with flour before sieving add sugar-salt solution and prepare dough.
2. Add tutee fruity after chopping at the last stage of mixing
3. Mould longer than brea.
4. Place it on greased tray. Proof till the desired volume achieved.
5. Title baked toast-loaf is allowed to cool until it acquires room temperature. Slice it and arrange each slice in baking sheet separately and dry in oven at 80°C to 90°C temperature for about 2 hours.

Observations

S.No.	Ingredients	Weight of Ingredients (gm)	%	Cost/kg (Rs.)	Cost/toast (Rs.)
1.	Flour	100	100		
2.	Shortening	5	5		
3.	Sugar	15	15		
4.	Yeast	2	2		
5.	Salt	2	2		
6.	Milk Powder	6	6		
7.	Tutti Fruity	10	10		
8.	Water	60	60		
9.	Liquid Glucose	3	3		

Weight of dough = ...g

Cost of dough = ...Rs.

Yield = ...g

Note your sensory observations below:

External	Internal
• Crust Colour	• Crumb Colour
• Volume Gain	• Symmetry of form
• Aroma	• Sensory

Calculations

1. Over Head Cost =	1. Baking-Cooling losses = Wt. of dough–Wt. of final product
2. Total cost = Ingredient cost + O.H.C.	2. Baking-Cooling loss (%)

Practical No. 12

Aim: Preparation of Mava Toast.

Procedure

1. Prepare dough by mixing the powdered ingredients with flour before sieving.
2. Add sesame seed at the last stage of mixing.
3. Mould longer than bread.
4. Place it in a greased tray. Proof till the desired volume achieved.
5. The baked toast-loaf is allowed to cool until it acquires room temperature. Slice it and arrange each slice on baking sheet separately and dry in oven at 80°C to 90°C temperature for about 2 hours.

Observations

S.No.	Ingredients	Weight of ingredients (gm)	%	Cost/kg (Rs.)	Cost/ toast (Rs.)
1.	Flour	100	100		
2.	Shortening	10	10		
3.	Sugar	25	25		
4.	Yeast	1.5	1.5		
5.	Salt	1	1		
6.	Water	60	60		
7.	Milk Powder	6	6		
8.	Oil	5	5		
9.	Tal	2	2		

Calculations

Weight of dough = ...gm

Cost of dough = ...Rs.

Yield = ...gm Or pieces

Note your sensory observations below:

External	Internal
• Crust Colour	• Crumb Colour
• Volume Gain	• Symmetry of form
• Aroma	• Sensory

Practical No. 13

Aim: Preparation of Pizza.

Procedure

Activation of Yeast

1. Take yeast in a bowl and add 1-2 tsp of lukewarm water and mix well. Add a pinch of sugar and 1 tsp of maida and mix it vigorously. Put this yeast suspension in a proover at 37°C and 70 per cent RH for 15 minutes.

2. Make a mixture of salt, skim milk powder and sugar in some amount of water and mix well.

3. Add improver to the remaining maida and sieve atleast twice.

Preparation of Pizza Dough and Base

1. Take a mixing bowl, add remaining water to it and salt, sugar and skim milk powder. Add half maida to it and yeast suspension and mix at a speed of 2. Add rest of the maida and mix at a speed of 2 then 4 and finally at a speed of 6. This will form pizza dough.

2. Grease the bowl lightly and put the dough in it. Now put the bowl in proover at 37°C and 70 per cent RH for 15 minutes.

For dividing and rounding

1. After 15 minutes take out the dough and cut in 4 equal parts, make the balls and cover it for 10 minutes. Start rolling the balls in a pizza base to get a diameter of 6 inches. Put this on a greased plate and keep it in a proover at 42°C and 80 per cent RH for 45 minutes.

2. Take out the bases and prick them with fork.

3. Bake at 220°C for 3-4 minutes and then turn and bake it for 2 minutes.

For Filling

1. Chop onion, cabbage, green chillies finely.

2. Chop garlic and grate ginger finely.

3. Boil the oil. Sauté it with cumin seed.

4. Add green chilli followed by onion and sauté till onion gets golden brown.

5. Add green masala, cover lid and allow cook for about 15-20 minutes.

6. Follow by salt and mixed hot spices.

Assembling (Base and Filling)

1. Assembling can be done before baking pizza base, spread base with tomato sauce followed by filling. Garnish with grated cheese, black pepper, tomato sauce.

Observations

S.No.	Ingredients	Weight of Ingredients (gm)	%	Cost/kg (Rs.)	Cost (Rs.)
1.	Refined Flour or Maida	150 g			
2.	Compressed Yeast	7g			
3.	Water	150ml			
4.	Salt	5g			
5.	Sugar	12g			
6.	Vegetable Oil	15g			
7.	Improver	2.5 g			
8.	Skimmed Milk Powder	5g			
For Pizza Topping					
9.	Ginger and Garlic (optional)	15g or as per taste			
10.	Onion	1 small (~50 g)			
11.	Capsicum	1 small (~50 g)			
12.	Tomato	1 small (~75 g)			
13.	Salt and Pepper	As per taste			
14.	Pizza Cheese or Mozzarella Cheese	To cover the topping properly or as per taste			

Weight of dough =   g

Cost of dough   =   Rs.

Yield   =   Pieces

Note your sensory observations below:

External	Internal
• Crust Colour	• Crumb Colour
• Volume Gain	• Symmetry of form
• Aroma	• Sensory

Calculations

1. Over Head Cost =	1. Baking-Cooling losses = Wt. of dough–Wt. of final product
2. Total cost = Ingredient cost + O.H.C.	2. Baking-Cooling loss (%)

Notes

1. Compressed yeast is used as a leavening agent.
2. After second proofing the pizza is pricked to prevent volume rise in base.
3. Vegetable oil is added at last because if it is added in beginning then it may bind gliadin and glutenin of flour and may prevent the formation of gluten network.

Practical No. 14

Aim: Preparation of Pitza.

Procedure

Pitza Base

1. Rub the shortening into flour and follow bread dough preparation procedure.
2. Rest the dough for 20 to 30 minutes.
3. Knock-back and rest for 15 min (optional).
4. Sheet the dough to 0.5 to 1.0 cm. Thickness and cut into round shape of about 15 to 18 cm diameter (use pitza cutter). The scrap dough is mixed and proofed for 5 to 10 min and resheeted.
5. Place on a baking tray and dock with a fork.
6. Allow to proof for 15 to 20 min and bake in such a way that bottom gets light brown colour and top surface remains almost white on slightly brownish.

For Filling

1. Chop onion, cabbage, green chillies finely.
2. Chop garlic and grate ginger finely.
3. Boil the oil. Sauté it with cumin seed.
4. Add green chilli followed by onion and sauté till onion gets golden brown.
5. Add green masala followed by cabbage and cover lid and allow cook for about 15-20 minutes.
6. Meanwhile, mix corn flour into a little water to prepare thick paste. Add corn flour solution into vegetables. Follow by salt and mixed hot spices.

Assembling (Base and Filling)

1. Assembling can be done before baking pitza base, spread base with tomato sauce followed by filling. Garnish with grated cheese, black pepper, tomato sauce.

Observations

S.No.	Ingredients	Weight of Ingredients (gm)	%	Cost/kg (Rs.)	Cost (Rs.)
1.	Flour	150			
2.	Shortening	30			
3.	Sugar	15			
4.	Yeast	5			
5.	Salt	3			
6.	Water	75-80			
7.	Cabbage	150			
8.	Onion	75			
9.	Green Chilli	25			
10.	Cardamom	25			
11.	Ginger	25			
12.	Salt	3			
13.	Turmeric	2			
14.	Clove Powder	1.5			
15.	Black Pepper	1.25			
16.	Cheese	20			
17.	Ajwain	1.25			
18.	Shortening	15			
19.	Corn Flour	5			
20.	Tomato Ketch-up	30			
	Total	**662 gm**			

Weight of dough = ...gm

Cost of dough   = ...Rs.

Yield   = ...Pieces

Note your sensory observations below:

External	Internal
• Crust Colour	• Crumb Colour
• Volume Gain	• Symmetry of Form
• Aroma	• Sensory

Calculations

1. Over Head Cost =	1. Baking-Cooling losses = Wt. of dough–Wt. of final product
2. Total cost = Ingredient cost + O.H.C.	2. Baking-Cooling loss (%)

CONFECTIONERY

Practical No. 1

Aim: Preparation of Nankhatai.

Recipe 1

Procedure

1. Add all the dry ingredients like baking powder, baking soda into flour and sieve to mix thoroughly twice and remove impurities.
2. Cream the vanaspati ghee till light and fluffy.
3. Add sugar gradually, continuing the creaming process, till the mixture become light, bright and fluffy.
4. Add milk in sufficient quantity (5 ml) and mix thoroughly.
5. Now add flour (already mixed with dry ingredients). Mix lightly and made into smooth dough. It is advisable to divide flour into 2 to 3 parts and add one after another.
6. Break the dough into small even size pieces of required size (20–25 gm).
7. Place in baking tray 1 cm (1/2") apart press gently and bake at 175°C (350°F) for 15 minutes.

Observations

S.No.	Ingredients	Weight of flour basis (gm)	%	Cost / kg (Rs.)	Cost (Rs.)
1	Flour	100 gm	100		
2	Vanaspati Ghee	60 gm	50-60		
3	Sugar	60 gm	60		
4	Soda	0.5 gm	0.5-2		
5	Elachi-Nutmeg Powder	2 gm	1-2		
6	Milk	5 ml	3-5		
	Total	**227.5gm**			

Yield   = ...gm
Weight of dough = ...gm
Cost of dough.   = ...Rs.
Note your sensory observations below:

External	Internal
• Crust Colour	• Crumb Colour
• Volume Gain	• Symmetry of form
• Aroma	• Sensory

Alternate Recipe-2

Procedure

1. Cream butter properly then add castor sugar to it.
2. Mix it properly and then churn it till a creamy and fluffy light mass of butter is formed. Add curd and little amount of cardamom powder and mix well. In the mean time dissolve soda in 2 teaspoon of water.
3. Cream the above butter well and then add soda to it and mix properly.
4. Add flour to it and make a dough with soft hand and give it a resting time of 10 minutes.
5. Make small ball and fatten them.
6. Bake at 180-190°C for 35 minutes till a slight pale colour appear on the crust.
7. Write the observations and do the calculations.

Observations

S.No.	Raw Materials	Weight of flour basis
1.	Flour / Maida	200 gm
2.	White Butter	190 gm
3.	Sugar	140 gm
4.	Soda	2 gm
5.	Cardamom Powder	5 gm
6.	Curd	10 gm
	Total	**227.5gm**

Calculations

1. Over Head Cost (OHC) =

2. Total or production cost = Ingredient cost + O.H.C

1. Baking/Cooling losses
   = Wt. of dough–Wt. of final product

2. Baking-Cooling loss (%)

Notes

1. They are the only type of cookies that are originated in India.
2. Cream the batter properly before adding soda water.
3. Do not flatter the balls more as it might not spread much.
4. Mix soda properly otherwise it will make holes on surface which will spoil the appearance.
5. Bake the cookies till the appearance of pale peach colour. Do not overbake them.

Practical No. 2

Aim: Preparation of Vanilla Biscuit.

Procedure

1. Sieve the flour for two times and mix all the dry ingredients like corn flour, custard powder and milk powder.
2. Cream the HVO or vanaspati in one direction till it becomes light and fluffy.
3. Add sugar gradually, continuing the creaming process, till the mixture gets light, bright and fluffy.
4. Add milk in sufficient quantity with sodium bicarbonate, ammonium bicarbonate and mix thoroughly.
5. Now add flour and mix lightly into smooth dough.
6. Roll out the dough into about (1/8th inch) thickness.
7. Prick with fork or design with grooved rolling pan.
8. Cut with desired shaped biscuit cutter.
9. Place on a baking tray 1 cm (1/2") apart from each other and bake.
10. Bake it at a temperature of 175°C (350°F) for about 10 to 12 minutes.

Observations

S.No.	Raw Materials	Weight of flour basis	%	Cost/Kg (Rs.)	Cost (Rs.)
1	Flour	100 gm	100		
2	Ghee	50 gm	50-60		
3	Sugar	50 gm	50		
4	Corn Flour	5 gm	5		
5	Custard Powder	5gm	5		
6	Milk Powder	5 gm	5		
7	Ammonium Bicarbonate	0.5 gm	0.5		
8	Sodium Bicarbonate	0.5 gm	0.5		
9	Vanilla Essence	0.5ml	0.5		
10	Milk or Water	20 ml	15-20		
	Total	**236.5**			

Weight of dough = 236.5 gm
Cost of Dough   = ...gm
Yield   = ...gm

Note your sensory observations below:

External	Internal
• Crust Colour	• Crumb Colour
• Volume Gain	• Symmetry of form

Calculations

1. Over Head Cost (OHC) =	1. Baking/ Cooling losses = Wt. of dough–Wt. of final product
2. Total or production cost = Ingredient cost + O.H.C.	2. Baking-Cooling loss (%)

Practical No. 3

Aim: Preparation of Chocolate Biscuit.
Procedure
Recipe - 1
1. Sieve the flour twice.
2. Remove little part of the flour and mix it with baking powder.
3. Cream the HVO or vanaspati in one direction till it gets fluffy and light.
4. Now add sugar gradually and continue the creaming process.
5. After this add milk, essence, salt and continue the creaming process.
6. Now add flour and mix lightly into smooth dough.
7. Now add baking powder along with the remaining flour into the dough and mix it again.
8. Make two parts of the dough and add coco powder into one part of it and mix it thoroughly.
9. Roll it into 3mm thick and cut it into even number of biscuit with a hand cutter having 5 cm diameter.
10. Now place it on baking tray and bake it at 175°C for 15 minutes.

Recipe - 2
1. In this method, make two parts of the dough and add one part with coco powder and make it into cylindrical shape.
2. Roll the dough and make it into small round balls and roll it like that for Nankhatai.
3. Now bake it at 175°C for 15 minute.
4. Cool it and pack it in polyethylene bag and store.

Observations

S.No.	Raw Materials	Weight of flour basis (gm)	%	Cost/kg. (Rs.)	Cost (Rs.)
1	Flour	100	100		
2	Ghee	60	60-70		
3	Sugar	50	50-60		
4	Baking Powder	0.5	0.5		
5	Cocoa Powder	5	5		
6	Salt	0.25	0.25		
7	Chocolate Essence	1	1		
8	Milk/Water	7.5	5-7.5		
9	Egg (Optional)	20	15.20		
	Total	**224.25**			

Weight of dough = 224.25 gm

Cost of Dough   = ...Rs.

Yield   = ...gm

Note your sensory observations below:

External	Internal
• Crust Colour	• Crumb Colour
• Volume Gain	• Symmetry of Form
• Aroma	• Sensory

Calculations

1. Over Head Cost (OHC) =	1. Baking/ Cooling losses = Wt. of dough–Wt. of final product
2. Total or production cost = Ingredient cost + O.H.C.	2. Baking-Cooling loss (%)

Practical No. 4

Aim: Preparation of Sweet and Salty Biscuits.

Procedure

1. Sieve the flour twice.
2. Remove little part of the flour and mix it with ammonium bicarbonate powder.
3. Cream the HVO/vanaspati in one direction till it becomes fluffy and light.
4. Now add sugar gradually and continue the creaming process.
5. After this add milk, salt and continue creaming
6. Now add flour and mix lightly into smooth dough.
7. Now add ammonium bicarbonate along with the remaining flour into the dough and then make it into smooth dough.
8. Now add ajwain or condiments into the flour and mix it.
9. Sheet the dough to about – 5mm thickness prick with fork, die with biscuit cutter and place them on tray for baking.
10. Bake the biscuit at 175°C for 15 to 20 minutes.
11. Cool the biscuit on cooling rack and then pack into polyethylene bags.
12. Store it at room temperature.

Observations

S.No.	Raw Materials	Weight of ingredient (gm)	%	Cost/ kg (Rs.)	Cost (Rs.)
1	Flour	100	100		
2	HVO/Margarine	50	40-50		
3	Sugar	15	15		
4	Ammonium Bicarbonate	1	1		
5	Salt	2.5	2.5		
6	Ajwain	2.5	-		
7	Milk	20	15-20		
	Total	**191 gm**			

Weight of dough = ...gm
Cost of Dough   = ...Rs.
Yield   = ...gm

Note your sensory observations below:

External	Internal
• Crust Colour	• Crumb Colour
• Volume Gain	• Symmetry of form
• Aroma	• Sensory

Calculations

1. Over Head Cost (OHC) =	1. Baking/ Cooling losses = Wt. of dough–Wt. of final product
2. Total or production cost = Ingredient cost + O.H.C.	2. Baking-Cooling loss (%)

Practical No. 5

Aim: Preparation of Melting Moments Cookies.

Procedure

1. Sieve the flour twice.
2. Now add baking powder in flour and again sieve the flour.
3. Cream the vegetable fat/HVO/Margarine and add grinded sugar slowly in small batches and continue the creaming process till it becomes light fluffy and bright coloured cream.
4. Add slowly with continuous mixing the flour being sieved.
5. At last cut the dough into 20 pieces of approximately same weight.
6. Roll the balls between the palms and make a ball.
7. Now add yellow colour to coconut shreds and mix it thoroughly.
8. After mixing, roll the balls with slight pressure to hold the coconut shreds.
9. Now place the balls into the baking tray and bake it at 175°C for about 15 minutes.
10. After baking, cool the product to room temperature.
11. After cooling, pack the product in polyethylene pouches of 200 gm each.

Observations

S.No.	Raw Materials	Weight of ingredient (gm)	%	Cost/ kg (Rs.)	Cost (Rs.)
1.	Flour	100	100		
2.	HVO/ Vanaspati	60	60-70		
3.	Sugar	60	50-60		
4.	Baking Powder	1	1		
5.	Coconut Essence	1	1		
6.	Coconut shreds	20	20		
7.	Milk	10	7-10		
8.	Yellow Colour	1	1		
9.	Salt	0.5	0-5		
	Total	**253.5**			

Weight of dough   = ...gm
Cost of dough   = ...Rs.
Yield   = ...gm

Note your sensory observations below:

External	Internal
• Crust Colour	• Crumb Colour
• Volume Gain	• Symmetry of Form
• Aroma	• Sensory

Calculations

1. Over Head Cost (OHC) =	1. Baking/ Cooling losses = Wt. of dough–Wt. of final product
2. Total or production cost = Ingredient cost + O.H.C.	2. Baking-Cooling loss (%)

Practical No. 6

Aim: Preparation of cherry short or tutti fruity biscuit.

Procedure

1. Sieve the flour atleast twice.
2. Cream the HVO/vanaspati in one direction till it become light and fluffy.
3. Now add grinded sugar slowly and in small parts with continuous mixing.
4. Now add the chopped cherry pieces and mix thoroughly and then add milk and sugar.
5. Now start adding flour slowly with mixing till all the flour is been accommodated into the cream.
6. After obtaining a smooth dough, roll it into a cylindrical rod shaped dough followed by cutting with dough cutter.
7. Make approximately 18-20 pieces and then roll it into sphere's and place in the baking tray 2" away from each other.
8. Now prick the biscuits with fork to give a designed effect.
9. Bake it at 170-175□C for about 12-15 minutes.
10. Cool the biscuits and then pack it and store it at room temperature.

Observations

S.No.	Ingredients	Weight of Ingredients (gm)	%	Cost/kg (Rs.)	Cost (Rs.)
1.	Flour	100	100		
2.	Shortening	60	60-70		
3.	Sugar	50	50		
4.	Cherry	15	15-20		
5.	Salt	0.25	0.15-0.25		
6.	Milk/Water	5	5		
Total		**230.25**			

Weight of dough = 230.25 gm
Cost of dough = ...Rs.
Yield = ...gm

Note your sensory observations below:

External	Internal
• Crust Colour	• Crumb Colour
• Volume Gain	• Symmetry of Form
• Aroma	• Sensory

Calculations

1. Over Head Cost (OHC) =	1. Baking/ Cooling losses = Wt. of dough–Wt. of final product
2. Total or production cost = Ingredient cost + O.H.C.	2. Baking-Cooling loss (%)

Practical No. 7

Aim: Preparation of Coconut Cookies.

Recipe 1

Procedure

1. Sieve the flour twice.
2. Cream the HVO/vanaspati in one direction till it becomes light and fluffy.
3. Continue creaming and add grounded sugar slowly in small batches.
4. Mix half of the shredded coconut with flour and add slowly into the cream of vanaspati and mix immediately.
5. Now at last add baking powder into the dough and mix it thoroughly into a smooth dough.
6. Roll the dough into a sheet of about 5 mm thickness. Now sprinkle the coconut shreds liberally and sheet it again to fix the coconut.
7. Die with the desired cut, place in a tray repeat the step No. 6 till all the dough is being over.
8. Bake it at 150°C for 15-20 minutes.
9. Cool the coconut Biscuit and then pack it into polyethylene bags and then store at room temperature

Observations

S.No.	Raw Materials	Weight of ingredient (gm)	%	Cost/Kg (Rs.)	Cost (Rs.)
1	Flour	100	100		
2	H.V.O.	60	60		
3	Sugar	50	50		
4	Baking Powder	1	0.5 to 1		
5	Coconut Essence	1	0.5 to 1		
6	Milk	15	15		
7	Coconut Shreds (desiccated)	30	30		
	Total	**257.0 gm**			

Weight of dough   =   257.0 gm
Cost of dough   = ...Rs.
Yield   = ...gm

Note your sensory observations below:

External	Internal
• Crust Colour	• Crumb Colour
• Volume Gain	• Symmetry of Form
• Aroma	• Sensory

Recipe 2
Procedure

1. Cream shortening with sugar using hand batter or whisk. Add egg in 2 equal parts and cream well. Add ammonium bicarbonate and sodium bicarbonate to it.
2. Add liquid to mixture also add some desiccated coconut to it.
3. Take out the mix and make dough with maida and baking powder with soft hand. Leave it for 10 minutes. Rolled the dough in thin sheets, spring desiccated coconut over it.
4. Stamp out the biscuits and bake at 200°C for 15-20 minutes.
5. Ensure that all biscuits should be of same size and thickness for proper baking.

S.No.	Raw Materials	Weight of ingredient (gm)	%	Cost/kg (Rs.)	Cost (Rs.)
1	Maida	100			
2	H.V.O. or Shortening	50			
3	Sugar	50			
4	Baking Powder	0.4			
5	Coconut Essence	Few drops			
6	Milk	15			
7	Desiccated Coconut Shreds	15- 20			
8	Egg	$1/3^{rd}$			
9	Ammonium Bicarbonate	0.5			
10	Sodium bicarbonate	0.4 to 0.5			
11	Liquid Glucose	2			
	Total				

Calculations

1. Over Head Cost (OHC) =	1. Baking/ Cooling losses = Wt. of dough–Wt. of final product
2. Total or production cost = Ingredient cost + O.H.C.	2. Baking-Cooling loss (%)

Note

1. In biscuits no increase in volume is required that's the reason why shortening is used instead of margarine.

2. Ammonium and sodium bicarbonate are added as they acts as leavening agent and also provide crispness.

3. To check the completeness of baking. The biscuit colour should be brown and should move on the tray on shaking. Also break the biscuit and check that uncooked dough at the center.

4. If the biscuits are not moving on the tray then reduce the temperature of oven and give more time.

5. If the biscuits are baked but colour has not developed then increase the temperature of the oven and leave it for 2-3 minutes.

Practical No. 8

Aim: Preparation of Falhari Nankhatai (for fasting).

Procedure

1. Replace flour with poppy seed flour, desiccated coconut and roasted dehulled finely crushed groundnut mixture and prepare dough by creaming method.
2. Made into small ball of walnut size, place in a baking tray and bake.

Observations

S.No.	Ingredients	Weight of ingredient (gm)	%	Cost / kg (Rs.)	Cost (Rs.)
1	Poppy Seed Flour	50	100		
2	Coconut Desiccated	50	100		
3	Groundnut	50	100		
4	Shortening	75	150		
5	Sugar	75	150		
6	Baking Powder	0.6	1		
7	Cardamom and Nutmeg	2.5	2/3		
8	Milk	10	5		
	Total	**313.01**			

Weight of dough　＝　313.01gm
Cost of dough　＝　.............Rs.
Yield　＝　.............gm

Note your sensory observations below:

External	Internal
• Crust Colour	• Crumb Colour
• Volume Gain	• Symmetry of Form
• Aroma	• Sensory

Calculations

1. Over Head Cost (OHC) =	1. Baking/ Cooling losses = Wt. of dough–Wt. of final product
2. Total or production cost = Ingredient cost + O.H.C.	2. Baking-Cooling loss (%)

Practical No. 9

Aim: Preparation of Orange Biscuit.

Procedure

1. Sieve the flour twice.
2. Now add baking powder, ammonium bicarbonate with small amount of flour.
3. Cream the vegetable ghee and add grounded sugar slowly in small batches and continue the creaming process till we get a light, fluffy and bright cream.
4. Add slowly with continuous mixing the custard powder and flour already being sieved.
5. Now add orange essence and orange colour and mix thoroughly and at last add baking powder and ammonium bicarbonate.
6. Sheet the dough it into about 5mm (1/6") thickness and then cut it with desired biscuit cutter.
7. Place it on baking tray and bake it at 150°C for about 20 minutes.
8. Cool the biscuits on cooling rack and then pack it in polyethylene bags and store it at room temperature.

Observations

S.No.	Ingredients	Weight of ingredients (gm)	%	Cost/ kg (Rs.)	Cost (Rs.)
1	Flour	100	100		
2	H.V.O.	60	60		
3	Sugar	50	50		
4	Custurd Powder	5	5		
5	Milk	10	10		
6	Baking Soda	1	1		
7	Ammonium Bicarbonate	1	1		
8	Orange Essence	1	1		
9	Colour	1	1		
	Total	**229.00**			

Weight of dough   = 229.00 gm
Cost of dough   = ...Rs.
Yield   = ...pieces

Note your sensory observations below:

External	Internal
• Crust Colour	• Crumb Colour
• Volume Gain	• Symmetry of Form
• Aroma	• Sensory

Calculations

1. Over Head Cost (OHC) =	1. Baking/ Cooling losses = Wt. of dough–Wt. of final product
2. Total or production cost = Ingredient cost + O.H.C.	2. Baking-Cooling loss (%)

Practical No. 10

Aim: Preparation of Square Nut Biscuit.

Procedure

1. Sieve the flour twice.
2. Add some flour to baking powder and sieve it.
3. Cream the HVO/Margarine and add grounded sugar in small batches and continue the creaming process till it becomes light, fluffy and bright cream.
4. Mix the roasted, dehulled, finely crushed groundnut with flour.
5. Now add essence and milk, slowly creaming and then start adding flour along with groundnut already mixed with it.
6. Make the dough into a square form and then place in refrigeration till it sets properly (about 1 hr.)
7. Slice with sharp knife or die into biscuit thickness (i.e., about 5 mm).
8. Apply egg/milk wash on top of the biscuits.
9. Place the biscuits into tray and bake them at 190°C for about 12 to 15 minutes.
10. Cool the biscuits on cooling rack and then pack and store it at room temperature.

Observations

S.No.	Ingredient	Weight of ingredient (gm)	%	Cost/ kg (Rs.)	Cost (Rs.)
1	Flour	150	100		
2	H.V.O.	90	60		
3	Sugar	75	50		
4	Baking Powder	0.5	0.5		
5	Vanila Essence	0.5	0.5		
6	Milk /Egg	20	16		
7	Groundnut	30	20		
	Total	**366**			

Weight of dough =   366gm
Cost of Dough   = ...Rs.
Yield   =..............pieces or gm

Note your sensory observations below:

External	Internal
• Crust Colour	• Crumb Colour
• Volume Gain	• Symmetry of Form
• Aroma	Sensory

Calculations

1. Over Head Cost (OHC) =	1. Baking/ Cooling losses = Wt. of dough–Wt. of final product
2. Total or production cost = Ingredient cost + O.H.C.	2. Baking-Cooling loss (%)

Practical No. 11

Aim: Preparation of Pineapple Cookie.

Procedure

1. Sieve the flour twice.
2. Add all the dry ingredients like baking powder, ammonia, custard powder, corn flour, milk powder, salt and mix it properly.
3. Cream the HVO/vanaspati till it becomes light and fluffy.
4. Add sugar gradually continuing the creaming process, till the mixture become light, bright and fluffy.
5. Mix the essence and colour.
6. Add milk in sufficient quantity and mix lightly.
7. Now add flour. Mix lightly and made into smooth dough. It is advisable to divide flour into 2 to 3 parts and add one after another.
8. Sheet it into about 5 mm thickness.
9. Now shape it with the desired biscuit cutter.
10. Now place it into baking tray and bake it at 150°C for about 15 to 20 minutes.
11. Now cool the biscuits and pack it and store at room temperature.

Observations

S.No.	Ingredients	Weight of Ingredient (gm)	%	Cost /kg (Rs.)	Cost (Rs.)
1	Flour	200	100		
2	H.V.O.	120	60		
3	Sugar	100	50		
4	Corn Flour	20	10		
5	Custard Powder	20	10		
6	Milk Powder	10	5		
7	Salt	1	0.5		
8	Baking Soda	0.5	0.25		
9	Ammonia Bicarbonate	0.5	0.25		
10	Pineapple Essence	0.5	0.25		
11	Yellow Colour	0.25	0.125		
12	Milk	20	10		
	Total	**492.75**			

Weight of dough = 492.75 gm

Cost of dough   = ...Rs.

Yield   = ...gm.

Note your sensory observations below:

External	Internal
• Crust Colour	• Crumb Colour
• Volume Gain	• Symmetry of Form
• Aroma	• Sensory

Calculations

1. Over Head Cost (OHC) =	1. Baking/Cooling losses = Wt. of dough–Wt. of final product
2. Total or production cost = Ingredient cost + O.H.C.	2. Baking-Cooling loss (%)

Practical No. 12

Aim: **Preparation of Peanut Cookie.**

Procedure

1. Roast and dehull the peanut and crush them finely.
2. Sieve the flour and mix the crushed peanuts with flour.
3. Cream the shortening till light and fluffy.
4. Add sugar gradually, continuing the creaming process till the mixture become light bright and fluffy.
5. Now add flour along with crushed ground nut along with baking powder after adding and properly mixing of milk into the cream.
6. At last add salt and baking powder and mix it till smooth dough is formed.
7. Allow the dough to relax in refrigerator for about 30 minutes.
8. Divide dough into small similar pieces and make it into ball.
9. Now place a peanut on each ball and press it lightly to fix it.
10. Place it on baking tray and bake it at 190°C for 12 to 15 minutes.
11. Cool the cookies on sieve and then pack it in polyethylene bag and store it at room temperature.

Observations

S.No.	Ingredients	Weight of Ingredient (gm)	%	Cost / kg (Rs.)	Cost (Rs.)
1	Flour	100	100		
2	Shortening	70	70		
3	Sugar	50	50		
4	Baking Powder	1.25	1.25		
5	Milk	10	10		
6	Salt	0.25	0.25		
7	Peanut	30	–		
	Total	**261.50**			

Weight of dough = 261.50 gm

Cost of dough = ...Rs.

Yield = ...gm

Note your sensory observations below:

External	Internal
• Crust Colour	• Crumb Colour
• Volume Gain	• Symmetry of Form
• Aroma	• Sensory

Calculations

1. Over Head Cost (OHC) =	1. Baking/ Cooling losses = Wt. of dough–Wt. of final product
2. Total or production cost = Ingredient cost + O.H.C.	2. Baking-Cooling loss (%)

Practical No. 13

Aim: Preparation of Cherry Cookie.

Procedure

1. Sieve the flour twice.
2. Add some flour to baking powder and sieve it.
3. Cream the vegetable ghee and add grounded sugar in small batches and continue the creaming process till it becomes light, fluffy and bright cream.
4. Mix the chopped cherry with flour.
5. Now add essence and milk, slowly creaming and then start adding flour along with cherry already mixed with it.
6. Make the dough into triangle shape and allow it to cool in refrigerator for 1 hour.
7. Slice with sharp knife into biscuit thickness [i.e. about 5 mm $(1/10^{th})$].
8. Now place the biscuits on the baking tray and bake it at 175°C for about 12 to 15 minutes.
9. Cool them and package and store it at room temperature.

Observations

S.No.	Ingredient	Weight of Ingredient (gm)	%	Cost / kg (Rs.)	Cost (Rs.)
1.	Flour	150	100		
2.	Shortening	90	60		
3.	Sugar	75	50		
4.	Cherry	25	10 to 25		
5.	Baking Soda	1.25	0.5 to 1		
6.	Salt	1.25	0.25 to 0.5		
7.	Essence Mix Fruit	1.25	0.5		
8.	Milk Powder	5	5 to 10		
9.	Milk	15	10 to 15		
	Total				

Weight of dough   = ...gm
Cost of dough   = ...Rs.
Yield   = ...gm

Note your sensory observations below:

External	Internal
• Crust Colour	• Crumb Colour
• Volume Gain	• Symmetry of Form
• Aroma	• Sensory

Calculations

1. Over Head Cost (OHC) =	1. Baking/ Cooling losses = Wt. of dough–Wt. of final product
2. Total or production cost = Ingredient cost + O.H.C.	2. Baking-Cooling loss (%)

Practical No. 14

AIM: Preparation of Jam Sandwiched Buns.

Procedure

1. Mix all the dry ingredients into flour and sieve twice.
2. Add shortening and rub thoroughly.
3. Add sugar and mix till it becomes light.
4. Finally add essence and mix it gently.
5. At last add the milk and make it into smooth dough.
6. Break the dough and make it into small balls.
7. Make a small depression and fill the jam.
8. Place them in baking tray and bake it at 190°C for about 15 to 20 minutes.
9. Cool the cookies and then pack in polyethylene bags.
10. Store cookies at room temperature.

Observations

S.No.	Ingredients	Weight of Ingredient (gm)	%	Cost / kg (Rs.)	Cost (Rs.)
1	Flour	100	100		
2	Shortening	50	50		
3	Sugar	50	50		
4	Baking Powder	1	0.2 to 1.25		
5	Essence	0.5	0.5		
6	Milk	20	15-20		
7	Jam	10	8 to 10		
	Total	**231.50**			

Weight of dough   =   231.50 gm

Cost of dough   = ...Rs.

Yield   = ...gm

Note your sensory observations below:

External	Internal
• Crust Colour	• Crumb Colour
• Volume Gain	• Symmetry of Form
• Aroma	• Sensory

Calculations

1. Over Head Cost (OHC) =

2. Total or production cost = Ingredient cost + O.H.C.

1. Baking/ Cooling losses
   = Wt. of dough–Wt. of final product

2. Baking-Cooling loss (%)

Practical No. 15

Aim: Preparation of Nut Ring Biscuit.

Procedure

1. Sieve the flour twice.
2. Now add milk powder to the flour.
3. Now mix the shortening after proper mixing with flour by rubbing method.
4. At last add the essence, salt, essence within the flour and shortening mixture.
5. At last add the sugar in it.
6. Now sheet the dough into 5 mm thickness and make into ring shape with dough nut cutter or die with big round biscuit cutter and make hole in the centre with a small cutter.
7. Brush the biscuit with milk and sprinkle freely with roasted, dehulled, crushed groundnut pieces and press gently to stick.
8. Place on tray and bake it at 175°C for about 12 to 15 minutes.

Observations

S.No.	Ingredients	Weight of ingredient (gm)	%	Cost / kg (Rs.)	Cost (Rs.)
1.	Flour	100	100		
2.	Ghee	60	60		
3.	Sugar	50	50		
4.	Milk	10	10		
5.	Salt	0.5	0.25 – 0.5		
6.	Vanilla Essence	0.5	0.5 to 1		
7.	Peanuts	25	20 to 25		
	Total	**236**			

Weight of dough　=　.............gm
Cost of dough　=　.............Rs.
Yield　　　　　=　.............gm

Note your sensory observations below:

External	**Internal**
• Crust Colour	• Crumb Colour
• Volume Gain	• Symmetry of Form
• Aroma	• Sensory

Calculations

1. Over Head Cost (OHC) =

2. Total or production cost = Ingredient cost + O.H.C.

1. Baking/ Cooling losses
   = Wt. of dough–Wt. of final product

2. Baking-Cooling loss (%)

Practical No. 16

Aim: Preparation of Falhari Biscuit.

Procedure

1. Roast, dehulled and crush the groundnut finely.
2. Cream the shortening till it become light and fluffy cream.
3. Add sugar gradually, continuing the creaming process, till the mixture become light, bright and fluffy.
4. Add milk in sufficient quantity (5ml) and mix thoroughly.
5. Now add dessicated coconut, groundnut.
6. After that add baking powder, cardamom nutmeg powder into it and mix evenly.
7. Break the dough into small even size pieces of required size.
8. Place in baking tray 1 cm (½″) apart press gently and bake at 175°C (350°F) for 15 minutes.

Observations

S.No.	Ingredients	Weight of Ingredients (g)	%	Cost / kg (Rs.)	Cost (Rs.)
1.	Groundnut	105	70		
2.	Dessicated Coconut	45	30		
3.	Shortening/HVO	40	25-30		
4.	Sugar	75	50		
5.	Cardamom-Nutmeg	1 to 1.5g	1		
6.	Milk	12ml	5 to 7		
7.	Baking Powder.	pinch	0.25		

Weight of dough = ...gm
Cost of dough = ...Rs.
Yield = ...gm

Note your sensory observations below:

External	Internal
• Crust Colour	• Crumb Colour
• Volume Gain	• Symmetry of Form
• Aroma	• Sensory

Calculations

1. Over Head Cost (OHC) =	1. Baking/ Cooling losses = Wt. of dough–Wt. of final product
2. Total or production cost = Ingredient cost + O.H.C.	2. Baking-Cooling loss (%)

Practical No. 17

Aim: Preparation of Coconut Castle.

Procedure

1. Add baking powder into flour and sieve twice.
2. Cream shortening and flour till light and fluffy as similar to sugar and fat.
3. Break egg in a bowl having smaller bottom than top and add essence. Place it on pot containing hot water. Be careful that the water does not touch to pot. Beat until stiff peak is obtained.
4. Add sugar gradually with continuous beating.
5. Remove from hot water as soon as all the sugar added.
6. Add flour-shortening mixture and blend thoroughly.
7. Grease and dust the tin and fill butter upto 3/4th portion.
8. Place on tray and bake.
9. Remove immediately from tin and allow it to cool.
10. Roll it out into jam-sugar syrup mix, followed by rolling in dessicated coconut.
11. Place it upside down and gently press half cherry on top.

Observations

S.No.	Ingredients	Weight of ingredients (gm)	%	Cost / kg (Rs.)	Cost (Rs.)
1.	Flour	150	100		
2.	Shortening	120	80		
3.	Sugar	120	180		
4.	Egg	3 no.	35-40		
5.	Baking Powder	½ tea spoon	1.5		
6.	Shredded Coconut	20	12-15		
7.	Salt	Pinch	0.25		
8.	Vanilla Essence	½ tea spoon	1-1.5		
9.	Coconut (Garnishing)	40	20-25		
10.	Jam	30	20-25		
11.	Sugar Syrup	40	30-35		
12.	Colour	1	0.5		
13.	Milk	25	15.20		
	Total	**516**			

Weight of dough   = ...gm
Cost of dough   = ...Rs.
Yield   = ...gm

Note your sensory observations below:

External	Internal
• Crust Colour	• Crumb Colour
• Volume Gain	• Symmetry of Form
• Aroma	• Sensory

Calculations

1. Over Head Cost (OHC) =	1. Baking/ Cooling losses = Wt. of dough–Wt. of final product
2. Total or production cost = Ingredient cost + O.H.C.	2. Baking-Cooling loss (%)

Practical No. 18

Aim: Preparation of Cinnamon Crescent.

Procedure

1. Mix coarse sugar and cinnamon powder.
2. Prepare dough by rubbing method.
3. Mix all the dry ingredients into flour and sieve twice.
4. Add shortening and rub thoroughly.
5. Add sugar and mix lightly.
6. Add essence into it. Save little egg for washing.
7. Place it on butter paper and sheet the dough into rectangular shape of 3mm (1/8") thickness.
8. Apply egg wash on half portion of the dough, sprinkle sugar-cinnamon mixture.
9. Fold the remaining half portion of dough over it and roll to biscuit thickness, if required. It will also help to fix the sugar.
10. Die with small round biscuit cutter and again into the crescent and leaf shape.
11. Place the dough pieces on tray and bake.

Observations

S.No.	Ingredients	Weight of ingredients (gm)	%	Cost / kg (Rs.)	Cost (Rs.)
1	Flour	100			
2	Shortening	60			
3	Sugar	50			
4	Egg	½ no			
5	Sugar and Cinnamon	7.5			
6	Baking Powder	1/8 teaspoon (0.629)			
7	Vanilla Essence	5ml			
	Total	**217.5**			

Weight of dough   = ...gm
Cost of dough   = ...Rs.
Yield   = ...gm

Note your sensory observations below:

External	Internal
• Crust Colour	• Crumb Colour
• Volume Gain	• Symmetry of Form
• Aroma	• Sensory

Calculations

1. Over Head Cost (OHC) =	1. Baking/ Cooling losses = Wt. of dough–Wt. of final product
2. Total or production cost = Ingredient cost + O.H.C.	2. Baking-Cooling loss (%)

Practical No. 19

Aim: Preparation of Tricolour Biscuits.

Tricolour biscuit are plain biscuit each with biscuit having holes with butter icing of different colour.

Procedure

1. Prepare dough by rubbing method.
2. Roll it into 3 mm thickness.
3. Cut even number of biscuit with cutter of about 7 cm diameter.
4. Prick half the biscuit with fork and make three holes in the remaining biscuit in such a way that it forms triangle.
5. Place it on baking tray and bake.
6. Allow to cool completely.
7. Divide Jam into three parts. Colour them differently. Fill each are into three holes.

Observations

S.No.	Ingredients	Weight of ingredient (gm)	%	Cost / kg (Rs.)	Cost (Rs.)
1.	Flour	18	100		
2.	Shortening	58	60		
3.	Sugar	18	50		
4.	Baking Powder	40	1		
5.	Milk	22	5		
6.	Jam	120	25		
7.	Icing Sugar	-	50		
8.	Ghee	58	25		
9.	Essence	400	0.6		
	Total	**834**			

Weight of dough   = ...gm
Cost of dough   = ...Rs.
Yield   = ...pieces

Note your sensory observations below:

External	Internal
• Crust Colour	• Crumb Colour
• Volume Gain	• Symmetry of Form
• Aroma	• Sensory

Calculations

1. Over Head Cost (OHC) =	1. Baking/ Cooling losses
	= Wt. of dough–Wt. of final product
2. Total or production cost = Ingredient cost + O.H.C.	2. Baking-Cooling loss (%)

Practical No. 20

Aim: Preparation of Masala Biscuit.

Procedure

1. Chop green chillies, and masala very fine and make into paste in a mixer jar.
2. Prepare a smooth dough as per the rubbing method.
3. Mix the green masala paste into fat rubbed flour for Masala biscuit.
4. Sheet the dough to about 5 mm thickness.
5. Prick with fork, die with biscuit cutter, place on tray and bake.

Observations

S.No.	Ingredients	Weight of Ingredients (gm)	%	Cost/kg (Rs.)	Cost (Rs.)
1.	Flour	150	100		
2.	Ghee	80	53		
3.	Sugar	18	12		
4.	Salt	3.75	2.5		
5.	Green Masala	15	10		
6.	Baking Powder	1.25	1		
	Total	**268gm**			

Weight of dough   = ...gm
Cost of dough   = ...Rs.
Yield   = ...pieces

Note your sensory observations below:

External	Internal
• Crust Colour	• Crumb Colour
• Volume Gain	• Symmetry of Form
• Aroma	• Sensory

Calculations

1. Over Head Cost (OHC) =	1. Baking/ Cooling losses = Wt. of dough–Wt. of final product
2. Total or production cost = Ingredient cost + O.H.C.	2. Baking-Cooling loss (%)

Experimental and Analytical Bakery

Practical No. 21

Aim: Preparation of atta (whole wheat flour) biscuit.

Procedure

1. Add milk powder into wheat flour.
2. Mix ammonia, soda and cardamom powder into separated wheat flour.
3. Prepare a dough by creaming method.
4. Add the separated flour in to dough.
5. Mix it properly after adding milk.
6. Sheet the dough about 5 mm thickness. Sprinkle vegetable oil on to it. Prick with fork and die with biscuit cutter, place on tray and bake.

Observations

S.No.	Ingredients	Weight of Ingredients (gm)	%	Cost/kg (Rs.)	Cost (Rs.)
1.	Wheat Flour	150	100		
2.	Shortening	90	60		
3.	Sugar	75	50		
4.	Milk Powder	10 (2TS)	6-7		
5.	Ammonia	0.6 (1/8TS)	0.4		
6.	Soda	0.6 1/6(TS)	0.4		
7.	Vegetable Oil	10	6-7		
8.	Milk	20	12-14		
9.	Cardamom Powder	0.6 (1/8 TS)	0.4		
	Total				

Weight of dough   = ...gm
Cost of dough   = ...Rs.
Yield   = ...gm

Note your sensory observations below:

External	Internal
• Crust Colour	• Crumb Colour
• Volume Gain	• Symmetry of Form
• Aroma	• Sensory

Calculations

1. Over Head Cost (OHC) =	1. Baking/ Cooling losses = Wt. of dough–Wt. of final product
2. Total or production cost = Ingredient cost + O.H.C.	2. Baking-Cooling loss (%)

Practical No. 22

Aim: Preparation of Sweet Biscuit.

Procedure

1. Add cornflour into wheat flour.
2. Separate little amount of flour and add soda and ammonia into it.
3. Prepare dough by creaming method.
4. Add little amount of essence and mix separated flour containing soda and ammonia.
5. Sheet the dough about 5 mm thickness prick with fork and die with biscuit cutter, place on tray and bake.

Observations

S.No.	Ingredients	Weight of ingredient (gm)	%	Cost / kg (Rs.)	Cost (Rs.)
1.	Flour	150	100		
2.	Shortening	90	60		
3.	Sugar	75	50		
4.	Custard Powder	10	6-7		
5.	Corn Flour	10	6-7		
6.	Soda	0.6 (1/8 TSP)	0.4		
7.	Ammonia	0.6 (1/8 TSP)	0.4		
8.	Milk	15	10		
9.	Essence	1.5 (1/2 TSP)	1-1.5		
	Total	—			

Weight of dough   = ...gm
Cost of dough   = ...Rs.
Yield   = ...gm

Note your sensory observations below:

External	Internal
• Crust Colour	• Crumb Colour
• Volume Gain	• Symmetry of Form
• Aroma	• Sensory

Calculations

1. Over Head Cost (OHC) =

2. Total or production cost = Ingredient cost + O.H.C.

1. Baking/ Cooling losses
= Wt. of dough–Wt. of final product

2. Baking-Cooling loss (%)

Practical No. 23

Aim: Preparation of Ester Biscuit.

Procedure

1. Chop tutti fruity into small pieces and sieve the flour twice.
2. Cream the shortening and sugar by using creaming method.
3. Now add small quantity of flour with continuous mixing.
4. Add the remaining flour and make it into smooth dough.
5. Add walnut pieces and then bake it.

Observations

S.No.	Ingredients	Weight of ingredient (gm)	%	Cost / kg (Rs.)	Cost (Rs.)
1.	Flour	100	100		
2.	Shortening	75	50		
3.	Sugar	75	50		
4.	Tutti Fruity	30	20		
5.	Yellow Colour	0.25	0.25		
6.	Milk Powder	15	10		
7.	Vanilla Essence	0.6	0.6		
8.	Milk	20	13-14		
9.	Ammonia	0.6	0.6		
10.	Baking Powder		0.6		
	Total	**367.7**			

Weight of dough   = ...gm
Cost of dough   = ...Rs.
Yield   = ...gm

Note your sensory observations below:

External	Internal
• Crust Colour	• Crumb Colour
• Volume Gain	• Symmetry of Form
• Aroma	• Sensory

Calculations

1. Over Head Cost (OHC) =	1. Baking/ Cooling losses = Wt. of dough–Wt. of final product
2. Total or production cost = Ingredient cost + O.H.C.	2. Baking-Cooling loss (%)

Practical No. 24

Aim: Preparation of Salty and Spicy Cashew Shaped Biscuit.

Procedure

1. Prepare dough by rubbing method.
2. Add sugar, salt, jeera, ajwain and baking powder.
3. Sheet the dough and cut it in cashew shape followed by baking.

Observations

S.No.	Ingredients	Weight of ingredient (gm)	%	Cost / kg (Rs.)	Cost (Rs.)
1.	Flour	100	100		
2.	Shortening	30	30		
3.	Sugar	15	15		
4.	Salt	2.5	2.5		
5.	Water	50	50		
6.	Jeera, Ajwain	2.5	2.5		
7.	Baking Powder	0.5	0.5		
8.	Shortening	30	25		
	Total	**230**			

Weight of dough = ...gm
Cost of dough = ...Rs.
Yield = ...gm

Note your sensory observations below:

External	Internal
• Crust Colour	• Crumb Colour
• Volume Gain	• Symmetry of Form
• Aroma	• Sensory

Calculations

1. Over Head Cost (OHC) =	1. Baking/ Cooling losses = Wt. of dough–Wt. of final product
2. Total or production cost = Ingredient cost + O.H.C.	2. Baking-Cooling loss (%)

Practical No. 25

Aim: Preparation of Vegetable Puff.

Procedure

1. Prepare dough using 50 per cent shortening.
2. Prepare filling as per the raw material used.
3. Boil potato in salt water. Allow to cool till become warm. Remove skin and mesh clean and chop or grate the green masala and garlic.
4. Heat oil in a pan and add cumin seed and allow to crackle.
5. Add onion and sauté until golden brown.
6. Add green masala, garlic and suite.
7. Add salt, mixed hot spices. Mix thoroughly cook if required, remove from fire and allow to cool.
8. Roll out the dough into even thickness of 0.3 to 0.4 cm.
9. Cut into square shape.
10. Divide the filling into equal parts and make round balls.
11. Place one round ball in the centre of each square piece. Egg/milk wash the 0.5 cm horizontal ends and turn both together and press gently.

Observations

S.No.	Ingredients	Weight of ingredients (gm)	%	Cost / kg (Rs.)	Cost (Rs.)
1.	Flour	150	100		
2.	Shortening	90	50		
3.	Salt	3.75	2.5		
4.	Water	90	60		
5.	Potato	100	50		
6.	Onion	50	25		
7.	Chilli	10	2.5		
8.	Ginger	10	2.5		
9.	Coriander Leaves	10	2.5		
10.	Salt	3	1.25		
11.	Garam Masala	1	2.5		
12.	Taj	1			
13.	Long	1			
14.	Cumin Seed/Jeera	5			
15.	Turmeric	2			
16.	Lemon	3 ml			
17.	Ghee	15		5	
18.	Pea	20			
	Total	**562.75**			

Weight of dough   = ...gm
Cost of dough   = ...Rs.
Yield   = ...pieces

Note your sensory observations below:

External	Internal
• Crust Colour	• Crumb Colour
• Volume Gain	• Symmetry of Form
• Aroma	• Sensory

Calculations

1. Over Head Cost (OHC) =	1. Baking/ Cooling losses = Wt. of dough–Wt. of final product
2. Total or production cost = Ingredient cost + O.H.C.	2. Baking-Cooling loss (%)

Practical No. 26

Aim: Preparation of medallin cake.

Procedure

1. Add baking powder into flour and sieve twice.
2. Cream shortening and flour till light and fluffy cream is obtained.
3. Break egg in a bowl having smaller bottom than top and add essence. Place it on container containing hot water.
4. Add sugar gradually with continuous beating.
5. Remove from hot water as soon as all the sugar added.
6. Add flour shortening mixture and blend thoroughly.
7. Grease and dust the medallins tin and fill batter upto 3/4th portion.
8. Place on tray and bake.
9. Remove immediately from tin and allow to cool.
10. Roll it out into jam-sugar syrup mixture followed by rolling in desiccated coconut.

Observations

S.No.	Ingredients	Weight of ingredient (gm)	%	Cost / kg (Rs.)	Cost (Rs.)
1.	Flour	120			
2.	Shortening	120			
3.	Sugar	120			
4.	Egg	160			
5.	Essence	2.5			
6.	Baking Powder	1.25			
7.	Shredded Coconut	30			
8.	Cherry	–			
9.	Sugar Syrup	20			
10.	Jam	15			
	Total	**588.75**			

Weight of dough   = ...gm
Cost of dough   = ...Rs.
Yield   = ...pieces

Note your sensory observations below:

External	Internal
• Crust Colour	• Crumb Colour
• Volume Gain	• Symmetry of Form
• Aroma	• Sensory

Calculations

1. Over Head Cost (OHC) =	1. Baking/ Cooling losses = Wt. of dough–Wt. of final product
2. Total or production cost = Ingredient cost + O.H.C.	2. Baking-Cooling loss (%)

Practical No. 27

Aim: Preparation of Queen Cake.

Procedure

1. Add baking powder into flour and sieve twice.
2. Cream shortening and flour till light and fluffy as similar to sugar and fat.
3. Break egg in a bowl having smaller bottom than top and add essence. Place it on pot containing hot water. Be careful that the water does not touch to pot. Beat until stiff.
4. Add sugar gradually with continuous beating.
5. Remove from hot water as soon as all the sugar added.
6. Add flour shortening mixture and blend thoroughly.
7. Grease the fat into patty cake tin, dust with flour, turn up side down to remove excess dusting.
8. Place spoonful batter on each compartment and bake.
9. Bake it at 190°C temperature for 20 to 25 minutes.
10. After baking place it on cooling rack.
11. Cool it and pack it.

Observations

S.No.	Ingredients	Weight of ingredient (gm)	%	Cost / kg (Rs.)	Cost (Rs.)
1.	Flour	18	100		
2.	Shortening	58	100		
3.	Sugar	18	100		
4.	Egg	2.50	80		
5.	Baking Powder	40	1.5-2		
6.	Vanilla Essence	400	1-2		
7.	Milk	22			
8.	Tutti Fruity	36	50		
	Total	**594.50**			

Weight of dough = ...gm
Cost of dough = ...Rs.
Yield = ...No

Note your sensory observations below:

External	Internal
• Crust Colour	• Crumb Colour
• Volume Gain	• Symmetry of Form
• Aroma	• Sensory

Calculations

1. Over Head Cost (OHC) =	1. Baking/ Cooling losses = Wt. of dough–Wt. of final product
2. Total or production cost = Ingredient cost + O.H.C.	2. Baking-Cooling loss (%)

Practical No. 28

Aim: Preparation of Cherry Cake.

Procedure

1. Add all the dry ingredients like baking powder, corn flour, salt into flour and sieve twice to mix thoroughly and remove impurities.
2. Cream shortening and flour till light and fluffy as similar to sugar and fat.
3. Break eggs in a bowl having smaller bottom than top and add cream of toster in little amount for better fluffiness.
4. Add sugar continuously with beating.
5. Remove from hot water as soon as all the sugar added.
6. Add flour shortening mixture and blend thoroughly.
7. Add chopped and clean cherry and mix it into mixture.
8. Pour it into cake tin, and bake.
9. Bake it at 190°C for 20 to 25 minutes.
10. After baking place it on cooling rack.
11. Cool it and pack it.

Observations

S.No.	Ingredients	Weight of ingredient (gm)	%	Cost / kg (Rs.)	Cost (Rs.)
1	Flour	110	100		
2	Shortening	80	72		
3	Sugar	80	72		
4	Salt	Pinch	–		
5	Milk	40	36		
6	Cream of Foster	Pinch	–		
7	Cherry	50	45		
8	Egg	80	72		
9	Baking Powder	1.25	1		
10	Corn Flour	15	13		
	Total	—			

Weight of dough   = ...gm
Cost of dough   = ...Rs.
Yield   = ...gm

Note your sensory observations below:

External	Internal
• Crust Colour	• Crumb Colour
• Volume Gain	• Symmetry of Form
• Aroma	• Sensory

Calculations

1. Over Head Cost (OHC) =

1. Baking/ Cooling losses
   = Wt. of dough–Wt. of final product

2. Total or production cost = Ingredient cost + O.H.C.

2. Baking-Cooling loss (%)

Practical No. 29

Aim: Preparation of Colour Pastry.

Procedure

1. Prepare the dough by creaming method.
2. Divide into two portions and again divide only 1 portion into 2 parts.
3. Add red and green colour in each quarter position and roll out into pencil shape.
4. Roll out the half portion into 7 to 8 cm (2.1/2 to 3") wide strip of 3 mm (1/8") thickness. Apply egg wash.
5. Place both the pencil (joint together) in the middle of the strip and apply egg wash.
6. Now fold the dough strip on pencil so as make a diagonal.
7. Allow to cool in refrigerator until it sets properly (about ½ hour).
8. Cut dough into 5 mm (1/6") thickness slices. Place in tray and bake.

Observations

S.No.	Ingredients	Weight of ingredient (gm)	%	Cost / kg (Rs.)	Cost (Rs.)
1.	Flour	150	100		
2.	Shortening	90	60		
3.	Sugar	75	50		
4.	Baking Powder	0.6	–		
5.	Vanilla Essence	0.6	–		
6.	Milk	7.5	5		
7.	Egg	20	13		
8.	Colour	2.5			
	Total	**346.20**			

Weight of dough   = ...gm
Cost of dough   = ...Rs.
Yield   = ...gm

Note your sensory observations below:

External	Internal
• Crust Colour	• Crumb Colour
• Volume Gain	• Symmetry of Form
• Aroma	• Sensory

Calculations

1. Over Head Cost (OHC) =

2. Total or production cost = Ingredient cost + O.H.C.

1. Baking/ Cooling losses
   = Wt. of dough–Wt. of final product

2. Baking-Cooling loss (%)

Practical No. 30

Aim: Preparation of Pineapple and Cherry Upside Down Cake.

Procedure

1. Cream fat and sugar and put it on the bottom of cake tin.
2. Place pineapple slices on that and keep half cherry at the centre of each slice and fill the remaining part with walnut.
3. Prepare the batter by creaming method pour it on pineapple slices and bake.
4. Allow to cool for 5 to 10 minutes after baking and then transfer from tin to cooling sack in such a way that it turns upside down.

Observations

S.No.	Ingredients	Weight of ingredient (gm)	%	Cost / kg (Rs.)	Cost (Rs.)
1.	Shortening	50	25		
2.	Whole Sugar	50	25		
3.	Pineapple Slice	50			
4.	Cherry	50	25		
5.	Walnut	50	25		
6.	Flour	200	100		
6.	Shortening	120	60		
7.	Sugar	150	75		
8.	Egg	2 No.	40		
9.	Baking Powder	1.25	0.6		
10.	Pineapple Essence	1.25	0.8		
11.	Cardamom & Nutmeg	1.25	0.6		
12.	Sugar Syrup	30	15		
	Total	**953.75**			

Weight of dough = ...gm

Cost of dough = ...Rs.

Yield = ...gm

Note your sensory observations below:

External	Internal
• Crust Colour	• Crumb Colour
• Volume Gain	• Symmetry of Form
• Aroma	• Sensory

Calculations

1. Over Head Cost (OHC) =	1. Baking/ Cooling losses = Wt. of dough–Wt. of final product
2. Total or production cost = Ingredient cost + O.H.C.	2. Baking-Cooling loss (%)

Practical No. 31

Aim: Preparation of Egg Less Cup Cake/Cake.

Procedure

1. Add all the dry ingredients like baking powder into small quantity of flour and sieve the rest flour twice and remove impurities.

2. Cream the HVO/vanaspati till it becomes light and fluffy.

3. Add sugar gradually, continuing the creaming process till the mixtures become light, bright and fluffy.

4. Now mix milk powder with flour and then add it into the cream into small batches with continuous mixing and also adding baking powder, essence, salt and tutti fruity in it.

5. Mix the butter sufficient enough to get a smooth consistency having a positive drop test result.

6. In case of cup cake place paper cup into cup cake mold and pour batter upto 3/4th part and then place in a tray.

7. In case of cake grease with fat and dust with flour the cake tin. Then make upside down to remove the excess flour.

8. Pour the batter in cake tin and make pit at the centre, and then place on baking tray.

9. Die with the desired cut place in a tray repeat the step No. 6 till all the dough is being over.

10. Bake it at 190°C for about 20-25 minutes.

11. After baking remove the cup cake and the cake from the mould, cool it and then pack and store it at room temperature.

Observations

S.No.	Ingredients	%	Weight of Ingredient (gm)	Cost /kg (Rs.)	Cost (Rs.)
1	Flour + Corn Flour	100	100		
2	H.V.O.	65	65		
3	Sugar	80	80		
4	Milk Powder	30	30		
5	Tutty - Fruity	25	25		
6	Essence	1	1		
7	Water	70-100	75		
8	Baking Powder	0-5	0-5		
	Total		**376.50 gm**		

Weight of dough   =   376.50 gm
Cost of dough   = ... Rs.
Yield   = ... pieces or gm

Note your sensory observations below:

External	Internal
• Crust Colour	• Crumb Colour
• Volume Gain	• Symmetry of Form
• Aroma	• Sensory

Calculations

1. Over Head Cost (OHC) =	1. Baking/ Cooling losses = Wt. of dough–Wt. of final product
2. Total or production cost = Ingredient cost + O.H.C.	2. Baking-Cooling loss (%)

Practical No. 32

Aim: Preparation of Date and Walnut Cake.

Procedure

1. Cut the dates vertically into fine slices. And soak in water till it becomes soft (about ½ hrs.).
2. Prepare cake batter by creaming method.
3. Mix soaked date slice, chopped walnut, cashew nut in the batter.
4. Pour batter in cake tin and bake.

Observations

S.No.	Ingredients	Weight of ingredient (gm)	%	Cost / kg (Rs.)	Cost (Rs.)
1.	Flour	120			
2.	Shortening	100			
3.	Sugar	120			
4.	Egg	2 no			
5.	Baking Powder	1.25			
6.	Vanilla Essence	1.25			
7.	Dates	300			
8.	Walnut	50			
9.	Soda	1.25			
10.	Salt	0.25			
11.	Water	75 ml			
	Total				

Weight of dough = ...gm
Cost of dough = ...Rs.
Yield = ...gm

Note your sensory observations below:

External	Internal
• Crust Colour	• Crumb Colour
• Volume Gain	• Symmetry of Form
• Aroma	• Sensory

Calculations

1. Over Head Cost (OHC) =

2. Total or production cost = Ingredient cost + O.H.C.

1. Baking/ Cooling losses
   = Wt. of dough–Wt. of final product

2. Baking-Cooling loss (%)

Practical No. 33

Aim: Preparation of Coconut Macroons.

Procedure

1. Separate the egg white and place in a deep dry, ceramic dish. The egg white can easily be separated by making a small hole in the egg-top.
2. Add cream of tarter and allow to wet at peak stage (about 10 to 15 minute). Whisk it thoroughly with fork.
3. Add sugar gradually and keep on whisking until the mixture becomes stable enough to stand upright without spreading.
4. Add essence, colour, liquid glucose and salt just before the whisking is about to finish. Fold in the coconut very gently so that aeration is not lost and mixture remains stable.
5. Made into rock shape, place on baking tray and bake.

A modified product can also be developed by adding roasted, dehulled, crushed peanut instead of coconut and by removing the colour.

Observations

S.No.	Ingredients	Weight of ingredient (gm)	%	Cost / kg (Rs.)	Cost (Rs.)
1.	Shredded Coconut	100	100		
2.	Egg	40	30		
3.	Sugar	100	100		
4.	Cream of Heater	Pinch	0.25		
5.	Yellow Colour	0.25 ml	0.25-5		
6.	Vanilla Essence	1	1-1.5		
	Total	—			

Weight of dough   = ...gm
Cost of dough   = ...Rs.
Yield   = ...gm

Note your sensory observations below:

External	Internal
• Crust Colour	• Crumb Colour
• Volume Gain	• Symmetry of Form
• Aroma	• Sensory

Calculations

1. Over Head Cost (OHC) =

2. Total or production cost = Ingredient cost + O.H.C.

1. Baking/ Cooling losses
   = Wt. of dough–Wt. of final product

2. Baking-Cooling loss (%)

Practical No. 34

Aim: Preparation of Brownie Cake.

Procedure

1. Prepare the batter by creaming method.
2. Mix liquid glucose, chocolate essence, walnut and coco powder.
3. Keep in tin and bake.

Observations

S.No.	Ingredients	Weight of ingredient (g)	%	Cost/kg (Rs.)	Cost (Rs.)
1.	Flour	300	100		
2.	Shortening	250	80-85		
3.	Sugar	200	60-65		
4.	Egg	120	60		
5.	Baking Powder	2.5 (1/2 ts)	0.6-0.8		
6.	Soda	2.5	0.6-0.8		
7.	M. Sugar Powder	30 g (2 TS)	10		
8.	Coco Powder	25	8-10		
9.	Liquid Glucose	25	8		
10.	Chocolate Essence	2.5	1		
11.	Walnut	100	35-40		
12.	Salt	0.25	0.25 – 0.5		
13.	Milk or Water	150	50		
	Total	**1210.25**			

Weight of dough = ...gm
Cost of dough = ...Rs.
Yield = ...gm

Note your sensory observations below:

External	Internal
• Crust Colour	• Crumb Colour
• Volume Gain	• Symmetry of Form
• Aroma	• Sensory

Calculations

1. Over Head Cost (OHC) =

2. Total or production cost = Ingredient cost + O.H.C.

1. Baking/ Cooling losses
   = Wt. of dough–Wt. of final product

2. Baking-Cooling loss (%)

Practical No. 35

Aim: Preparation of Cherry Knob.

Procedure

1. Prepare the dough by creaming method.
2. Beat the egg yolk properly.
3. Roast the peanut and make small pieces of it.
4. Break the dough to the walnut size and roll into smooth ball.
5. Wash with milk and roll into groundnut pieces.
6. Place it on baking tray and press half cherry on top and bake.

Observations

S.No.	Ingredients	Weight of ingredient (gm)	%	Cost / kg (Rs.)	Cost (Rs.)
1.	Flour	150	100		
2.	Shortening	75	50		
3.	Sugar	75	50		
4.	Egg Yolk	15 (TS)	10		
5.	Baking Powder	1.5 (1/4 tsp)	1-1.5		
6.	Mix Fruit Essence	0.6 (1/2 tsp)	0.4		
7.	Peanut	50	33 – 34		
8.	Cherry	20	13-14		
9.	Milk	15	10		
	Total	—			

Weight of dough   = ...gm
Cost of dough   = ...Rs.
Yield   = ...gm

Note your sensory observations below:

External	Internal
• Crust Colour	• Crumb Colour
• Volume Gain	• Symmetry of Form
• Aroma	• Sensory

Calculations

1. Over Head Cost (OHC) =	1. Baking/ Cooling losses = Wt. of dough–Wt. of final product
2. Total or production cost = Ingredient cost + O.H.C.	2. Baking-Cooling loss (%)

Practical No. 36

Aim: Preparation of Dry Fruit Cake.

Procedure

1. Wash the dry fruits and whip with dry cloth.
2. Dip into caramel for 24 hours.
3. Add lemon juice, cardamom – nutmeg powder.
4. Prepare a batter by creaming method add caramel, liquid glucose, mix fruit essence and fruit at the last stage of mixing.
5. Place in suitable size cake tin and bake.

Observations

S.No.	Ingredients	Weight of ingredient (gm)	%	Cost / kg (Rs.)	Cost (Rs.)
1.	Flour	120	100		
2.	Shortening	80	65 – 70		
3.	Sugar	100	80 – 85		
4.	Egg	80 (2 no.)	65-70		
5.	B. Powder	2.5	1.5-2		
6.	Soda	0.6	0.4		
7.	Milk Powder	15	10-12		
8.	Corn Flour	15	10-12		
9.	Dry Fruit				
10.	Cashew Nut	50			
11.	Almond	50			
12.	Grapes	50			
13.	Tutti Frutti	50			
14.	Cherry	50	Pinch		
15.	Salt	0.25			
16.	Water or Milk	30	2		
17.	Mix Fruit Essence	2-5	20 – 25		
18.	Caramel	30	1.5-2		
19.	Cardamom-nutmeg	2.5	2 - 3		
20.	Lemon Juice	5.0			
	Total				

Weight of dough   = ...gm
Cost of dough   = ...Rs.
Yield   = ...gm

Note your sensory observations below:

External	Internal
• Crust Colour	• Crumb Colour
• Volume Gain	• Symmetry of Form
• Aroma	• Sensory

Calculations

1. Over Head Cost (OHC) =	1. Baking/ Cooling losses = Wt. of dough–Wt. of final product
2. Total or production cost = Ingredient cost + O.H.C.	2. Baking-Cooling loss (%)

Practical No. 37

Aim: Preparation of Welse Cheese Cake.

Procedure

Part - 1

1. Prepare biscuit dough by rubbing method.
2. Roll it out into 3 mm (1/8") thickness and round cut with biscuit cutter of about 7cm (3") round.
3. Place it on greased dusted petty cake tin and spread a little jam.

Part - 2

1. Prepare cake butter.
2. Place tea spoonful of butter on each biscuit dough piece and bake.

Observations

S.No.	Ingredients	Weight of ingredients (gm)	%	Cost / kg (Rs.)	Cost (Rs.)
1.	Flour	250	100		
2.	Shortening	90	35-40		
3.	Sugar	100	40		
4.	Baking Powder	3.75	1.5		
5.	Milk	90	35-40		
6.	Salt	Pinch	0.25		
7.	Soda	2.5	1-1.2		
8.	Ginger	10	4-5		
9.	Egg	80	30-35		
10.	Date	200	70-80		
11.	Vanilla Essence	2.5	1		
12.	Molasses (Caramel)	100	40		
	Total				

Weight of dough   = ...gm

Cost of dough   = ...Rs.

Yield   = ...gm

Note your sensory observations below:

External	Internal
• Crust Colour	• Crumb Colour
• Volume Gain	• Symmetry of Form
• Aroma	• Sensory

Calculations

1. Over Head Cost (OHC) =	1. Baking/ Cooling losses = Wt. of dough–Wt. of final product
2. Total or production cost = Ingredient cost + O.H.C.	2. Baking-Cooling loss (%)

Practical No. 38

Aim: Preparation of Date and Ginger Cake. ·

Procedure

1. Clean the date with water cut vertically into fine slices and soak into warm tea decoction/molasses and baking soda, keep it aside till it becomes soft for about 2 to 4 hour.
2. Mix ginger powder into flour before sieving.
3. Prepare cake butter by creaming method. Finally mix soaked date slice.
4. Pour the batter into cake tin and bake.

Observations

S.No.	Ingredients	Weight of ingredients (gm)	%	Cost/kg (Rs.)	Cost (Rs.)
1	Flour	250	100		
2	Shortening	90	35-40		
3	Sugar	100	40		
4	Baking Powder	3.75	1.5		
5	Milk	90	35-40		
6	Salt	Pinch	0.25		
7	Soda	2.5	1-1.2		
8	Ginger	10	4-5		
9	Egg	80	30-35		
10	Date	200	70-80		
11	Vanilla Essence	2.5	1		
12	Molasses (Caramel)	100	40		
	Total	**928.75**			

Weight of dough = ...gm

Cost of dough = ...Rs.

Yield = ...gm

Note your sensory observations below:

External	Internal
• Crust Colour	• Crumb Colour
• Volume Gain	• Symmetry of Form
• Aroma	• Sensory

Calculations

1. Over Head Cost (OHC) =	1. Baking/ Cooling losses = Wt. of dough–Wt. of final product
2. Total or production cost = Ingredient cost + O.H.C.	2. Baking-Cooling loss (%)

Practical No. 39

Aim: Preparation of Cream Rolls.

Procedure

1. Prepare hard dough, sheet it vertically.
2. Now cut along the dough into 5 equal pieces.
3. Now roll it onto the mould and then seal the upper side.
4. Bake it and allow it to cool.
5. Again keep it for evaporation /second baking for water.
6. Cream the shortening along with sugar into a smooth icing.
7. Now fill the rolls with cream with the help of cones.
8. Now spread the shredded coconut on top of it.

Observations

S.No.	Ingredients	Weight of ingredients (gm)	%	Cost / kg (Rs.)	Cost (Rs.)
1	Flour	150	100		
2	Shortening	10	5		
3	Shortening	40	35-40		
4	Salt	3.79	2.5-3		
5	Water	90	60		
6	Shortening	90	50-60		
7	Icing Sugar	115	75		
8	Essence	1.5	1		
9	Colour	0.5	0.25		
10	Shredded Coconut	7.5	3-5		
	Total	**508**			

Weight of dough = ...gm

Cost of dough = ...Rs.

Yield = ...No.'s

Note your sensory observations below:

External	Internal
• Crust Colour	• Crumb Colour
• Volume Gain	• Symmetry of Form
• Aroma	• Sensory

Calculations

1. Over Head Cost (OHC) =

2. Total or production cost = Ingredient cost + O.H.C.

1. Baking/ Cooling losses
   = Wt. of dough–Wt. of final product

2. Baking-Cooling loss (%)

Practical No. 40

Aim: Preparation of Coffee Buns.

Procedure

1. Prepare dough by rubbing method.
2. Divide into 50 gm pieces and round it.
3. Apply egg wash and roll the balls into sugar granules.
4. Place on baking tray and bake.

Observations

S. No.	Ingredients	Weight of ingredient (gm)	%	Cost / kg (Rs.)	Cost (Rs.)
1	Flour	150	100		
2	Shortening	75	50		
3	Sugar	75	50		
4	Baking Powder	1.25	1		
5	Salt	Pinch	-		
6	Egg	1 No.	25-30		
7	Chocolate Essence	1.25	1		
8	Coffee Powder	15	10		
9	Sugar (Whole)	30	20		
	Total	**387**			

Weight of dough   = ...gm
Cost of dough   = ...Rs.
Yield   = ...gm

Note your sensory observations below:

External	Internal
• Crust Colour	• Crumb Colour
• Volume Gain	• Symmetry of Form
• Aroma	• Sensory

Calculations

1. Over Head Cost (OHC) =	1. Baking/ Cooling losses = Wt. of dough–Wt. of final product
2. Total or production cost = Ingredient cost + O.H.C.	2. Baking-Cooling loss (%)

Practical No. 41

Aim: Preparation of Chocolate.

Procedure

1. Take sugar and make sugar syrup.
2. Then add butter into it followed with addition of milk powder.
3. Then mix the coco powder.
4. Take appropriate size of plate and grease it with oil.
5. Add the mixture, layer it properly, allow it to cool.
6. Put it into refrigeration.
7. Cut it into appropriate size.

Observations

S.No.	Ingredients	Weight of ingredient (gm)	%	Cost / kg (Rs.)	Cost (Rs.)
1.	Sugar	125			
2.	Milk Powder	100			
3.	Coco Powder	50			
4.	Butter	35			
	Total	**310**			

Weight of dough = ...gm
Cost of dough = ...Rs.
Yield = ...gm

Note your sensory observations below:

External	Internal
• Crust Colour	• Crumb Colour
• Volume Gain	• Symmetry of Form
• Aroma	• Sensory

Calculations

1. Over Head Cost (OHC) =	1. Baking/ Cooling losses = Wt. of dough–Wt. of final product
2. Total or production cost = Ingredient cost + O.H.C.	2. Baking-Cooling loss (%)

Practical No. 42

Aim: Preparation of Sponge Cake with Icing.

Procedure

1. Add all the day ingredients like baking powder, corn flour, milk powder, salt into flour and sieve twice to mix thoroughly and remove impurities.
2. Cream the HVO/vanaspati till it becomes light and fluffy.
3. Add sugar gradually during whipping of egg.
4. Mix egg, sugar mixer and flour alternatively.
5. Add milk or water in sufficient quantity and mix slightly in the batter. Add essence and colour.
6. Pour the batter in greased and dusted cake tin and bake.

For preparing icing

1. Cream shortening and sieved icing sugar gradually.
2. Add essence.
3. Divide the icing as required and mix colour thoroughly.

Observations

S.No.	Ingredients	Weight of ingredient (gm)	%	Cost / kg (Rs.)	Cost (Rs.)
1	Flour	80	100		
2	Shortening	120	65-70		
3	Sugar	10	100		
4	Corn Flour	12	10		
5	Milk Powder	80	10		
6	Egg	2.5	65-70		
7	Baking Powder	7.5	2		
8	Essence	Pinch	2		
9	Salt	25	0.25		
10	Milk/Water	120	20-25		
11	Icing sugar	120	150-200		
12	Shortening	1.25	90-100		
13	Essence	2	1		
14	Colour		1.5-2		
	Total	**765.25**			

Weight of cake = ...gm
Cost of cake   = ...Rs.

Note your sensory observations below:

External	Internal
• Crust Colour	• Crumb Colour
• Volume Gain	• Symmetry of Form
• Aroma	• Sensory

Calculations

1. Over Head Cost (OHC) =	1. Baking/ Cooling losses = Wt. of dough–Wt. of final product
2. Total or production cost = Ingredient cost + O.H.C.	2. Baking-Cooling loss (%)

Practical No. 43

Aim: Preparation of Chocolate Fudge.

Procedure

1. Mix sugar, condensed milk, cocoa, glucose, half quantity of ghee in a thick bottomed pan and heat until all sugar gets dissolved.
2. Now boil it about 1 hour on slow fire with constant stirring.
3. Once the mixture starts to form small ball, then remove from fire.
4. Immediately add essence and mix thoroughly.
5. Pour it into greased baking tray and spread evenly.
6. After few minutes mark to cut into desired pieces. Place in refrigerator make into pieces.

Recipe

S.No.	Ingredients	Weight of ingredients
1.	Condensed milk	400 gm
2.	Sugar (Granulated)	150 gm
3.	Liquid Glucose	1 TS
4.	Coco Powder	2 TS
5.	Shortening	80 gm
6.	Chocolate Essence	¼ teaspoon
7.	Water	400 ml
8.	Shreded coconut	30 gm

Practical No. 44

Aim: Preparation of Icing and Fondant.
- **Butter Icing**

Procedure

1. Sieve icing sugar to remove any lumps and to aerate it. If milk powder is used, sieve it with icing sugar.
2. Cream butter in the bowl till it become smooth and free from any lumps.
3. Add Glucose, condensed milk, glycerine and mix thoroughly by creaming.
4. Add icing sugar gradually and continue to cream until all sugar is added.
5. Now add emulsifies and cream till fat become light and fluffy.
6. Dissolve salt in water. Mix with acetic acid, essence and mix to fat.

Recipe:

S.No.	Ingredients	Weight of ingredient (gm)	%
1	Icing Sugar	100	100
2	Shortening	60	60
3	Essence	0.50 ml	5/1

- **Royal icing**

Procedure

1. Sieve icing sugar.
2. Separate egg white. Add lemon juice, essence and colour and beat lightly.
3. Add icing sugar gradually with continuous creaming until it acquires spreading consistency.
4. Mix glycerine, which prevent icing to become hard after setting.

Recipe:

S.No.	Ingredients	Weight of ingredient (gm)	%
1.	Icing Sugar	400	100
2.	Egg White	60	15
3.	Lemon Juice	1.25	0.5
4.	Cream of Tarter	1	0.5
5.	Essence	1.25	5/1
6.	Colour		If needed

• **Gum paste**

Procedure

1. Soak gelatin in water, heat by double boiler method and stir till the gelatin is dissolved completely.
2. Add glycerine and mix thoroughly.
3. Add icing sugar gradually and mix upto the dough consistency. Add corn flour if required.
4. Add colour and essence as required.

Recipe:

S.No.	Ingredients	Weight of ingredient (gm)	%
1.	Icing Sugar	400	100
2.	Gelatin	30	7
3.	Warm Water	45	10
4.	Essence	2.25	5
5.	Colour	–	If needed

• **Fondant**

Procedure

1. Sieve icing sugar. Add into water. Add liquid glucose. Heat at low temperature.
2. After boiling add lemon juice.
3. It becomes dense at 200°C.
4. Put off from gas.
5. Cream it till it gain shinning surface.

Recipe:

S.No.	Ingredients	Weight of ingredient (gm)	%
1.	Sugar	100	100
2.	Water	30	20
3.	Liquid glucose	10	10
4.	Citrus/Lemon Juice	1	5/1
5.	Essence	1	5/1
6.	Colour	–	If needed

• **Almond icing**

Procedure

1. Sieve icing sugar.
2. Separate egg white.
3. Add icing sugar gradually with continuous creaming.

4. Mix glycerine.

Recipe:

S.No.	Ingredients	Weight of ingredient (gm)	%
1.	Icing Sugar	250	100
2.	Almond Powder	250	100
3.	Egg White	40	16
4.	Citrus Juice	1.25	5/1
5.	Vanilla Essence	1.25	If needed
6.	Colour	–	If needed

Practical No. 45

Aim: Preparation of Plain Butter Sponge Cake.

Procedure

1. Cream the margarine with sugar till a fluffy mix is formed.
2. Break the eggs in a bowl and beat them properly. Take 2-3 teaspoon of liquid egg to another bowl and add cake gel to it, mix it thoroughly to a stiff peak stage. To the rest of the egg add appropriate amount of both the essence.
3. Add the egg plus cake gel mixture to the creamy mixture and mix it thoroughly using the kitchen aid machine and then add the egg with essence.
4. Sieve the maida with baking powder thoroughly atleast thrice.
5. Add this sieved maida to the above creamy mixture in three proportions by cut and fold method.
6. Pour this mixture to lined or greased tin and bake it a temperature of 180°C for 25 to 30 minutes.
7. Check the doneness of cake by inserting a toothpick. The toothpick should come out clean and be free from crumb particles.
8. After baking, take out the cake from the cake tin and allow it to cool.
9. After cooling, slice the cake and serve.

Observations

Ingredients	Weight of ingredients (gm)	%	Cost/kg (Rs.)	Cost (Rs.)
Maida	100			
Margarine/ HVO	80			
Sugar	100			
Egg	90 (~2 egg)			
Baking Powder	2			
Cake Gel	3 to 3.5			
Vanilla Essence	Few drops			
Lemon Essence	Few drops			

Notes:

1. Maida and baking powder should be sieved thrice to get uniform mixing and proper aeration.
2. Add sugar gradually (spoon by spoon) in the cream.
3. Proper care should be taken to ensure that cream doesn't become too flowy or runny.
4. Eggs should be kept at room temperature, before using in the cake preparation.

5. After the cake mixture is poured in the greased tin, do not keep the tin for long time as CO_2 may escape and this will not give honeycomb structure to the cake.
6. On crust, maillard browning and caramelization occur which provide brown colour to the crust.
7. Cake gel acts as emulsifier and stabilizer.

Note your sensory observations below:

External	Internal
• Crust Colour	• Crumb Colour
• Volume Gain	• Symmetry of Form
• Aroma	• Sensory

Calculations

1. Over Head Cost (OHC) =	1. Baking/ Cooling losses = Wt. of dough–Wt. of final product
2. Total or production cost = Ingredient cost + O.H.C.	2. Baking-Cooling loss (%)

Practical No. 46

Aim: Preparation of Chocolate Cake.

1. Cream the margarine with sugar till a fluffy mixture is formed and all the sugar has dissolved in the margarine.
2. Break the eggs in a bowl and beat them with the help of a spoon. Take 2-3 teaspoon of liquid egg to another bowl and add cake gel to it, mix it thoroughly to a stiff peak stage. To the rest of the egg add appropriate amount of both the essence.
3. Add the egg plus cake gel mixture to the creamy mixture and mix it thoroughly using the kitchen aid machine and then add the egg with added essence.
4. Sieve the maida, cocoa powder and baking powder atleast thrice for proper mixing.
5. Add this sieved maida mix to the creamy mixture in three to four proportions by cut and fold method so as to ensure proper mixing.
6. Pour this mixture to lined or greased tin and bake it at temperature of 180°C for 30 to 35 minutes.
7. Check the doneness of cake by inserting a toothpick. The toothpick should come out clean and be free from crumb particles.
8. After baking, take out the cake from the cake tin and allow it to cool.
9. After cooling, slice the cake and serve.

Observations

Ingredients	Weight of ingredients (gm)	%	Cost/kg (Rs.)	Cost (Rs.)
Maida	85			
Cocoa Powder	15			
Margarine/ HVO	80			
Sugar	105			
Egg	95 (~2 egg)			
Baking Powder	2			
Cake Gel	3 to 3.5			
Lemon Essence	2-4 drops			
Chocolate Essence	4-6 drops			

Notes:

1. Cocoa powder provides taste and colour to the cake. As it has drying effect on the cake so to overcome this effect the amount of egg is increased.
2. If after adding maida in the creamy mixture, the batter seems to be dry then add some water and mix it properly with the dough.
3. Add adequate or prescribed amount of the baking powder. As if too much of baking powder is added then cake will collapse and if amount is less then the cake will be compact.

Practical No. 47

Aim: Preparation of Sponge Cake with Pineapple Icing.

Procedure

I For Baking:

1. Break the eggs in a bowl and start beating them. After beating them for sometimes add castor sugar to it in parts after every 2 minutes, i.e., sugar is added gradually.

2. After half of the sugar is added, add egg and cake gel mixture to it.

3. Add the remaining sugar gradually and mix it by whisker till soft peak is observed in the batter.

4. Add the mixture of water, salt and essence in the batter in two parts, mix it thoroughly. Sieve maida and baking powder 3-4 times for proper mixing.

5. Add maida and baking powder mixture to the batter in three parts and mix it by using whisker.

6. Add vegetable oil to it and mix it at a speed of 2.

7. Pour the batter into a greased cake tin and bake it at a temperature of 200°C for 20-25 minutes.

8. Check the doneness of baking using a tooth pick. It should come out clean i.e. it should be free from crumb particles.

9. Take out the cake and allow it to cool immediately for icing.

II For Syrup:

1. Mix all the ingredients.

2. Once the cake has cooled sufficiently, then cut the cake horizontally from the center taking care that cake is cut in two equal halves.

3. Remove extra crumb from the surface. Apply syrup on both the surface of cake.

III For Icing:

1. Take 100gm cream and add icing sugar to it and beat it over ice bath till strong peak is seen. To check this stage- take the cream in a spoon, drop the cream if it doesn't fall i.e. it forms the peak then it is considered that cream is made.

2. Put some icing mix on the cut and moistened cake surface.

3. Put the other layer on it and add rest of the cream over it. Level the icing and smoothen it.

4. Garnish the cake with pineapple and cherries.

Observations

S.No.	Ingredients	Weight of ingredients (gm)	%	Cost/kg (Rs.)	Cost (Rs.)
I.	**For Baking**				
1.	Maida	100			
2.	Vegetable Oil	10 ml			
3.	Sugar	110			
4.	Egg	135 (~3 egg)			
5.	Baking Powder	2			
6.	Cake Gel	5			
7.	Salt	1			
8.	Water	35ml			
9.	Lemon and Pineapple Essence	Few drops			
II.	**For Icing**				
1.	Cream	100			
2.	Icing Sugar	50			
3.	Pineapple and Cherries	For garnishing			
III.	**For Syrup**				
1.	Water	100ml			
2.	Sugar	15-20			
3.	Vanilla/Pineapple Essence	3-4 drops			

Note your sensory observations below:

External	Internal
• Crust Colour	• Crumb Colour
• Volume Gain	• Symmetry of Form
• Aroma	• Sensory

Calculations

1. Over Head Cost (OHC) =

2. Total or production cost = Ingredient cost + O.H.C.

1. Baking/ Cooling losses
   = Wt. of dough–Wt. of final product

2. Baking-Cooling loss (%)

Practical No. 48

Aim: Preparation of Chocolate Sponge Cake with Icing.

Procedure

I For Baking:

1. Break the eggs in a bowl and start beating them. After beating them for sometimes add castor sugar to it in parts after every 2 minutes i.e. sugar is added gradually.

2. After half of the sugar is added, add egg and cake gel mixture to it.

3. Add the remaining sugar gradually and mix it by whisker till soft peak is observed in the batter.

4. Add the mixture of water, salt and essence in the batter in two parts, mix it thoroughly.

5. Sieve maida, cocoa powder and baking powder 3-4 times for proper mixing.

6. Add above mix of maida, cocoa powder and baking powder to the batter in three parts and mix it by using whisker.

7. Add vegetable oil to it and mix it at a speed of 2.

8. Pour the batter into a greased cake tin and bake it at a temperature of 200°C for 20-25 minutes.

9. Check the doneness of baking using a tooth pick. It should come out clean, i.e., it should be free from crumb particles.

10. Take out the cake and allow it to cool immediately for icing.

II For Syrup:

1. Mix all the ingredients.

2. Once the cake has cooled sufficiently, then cut the cake horizontally from the center taking care that cake is cut in two equal halves. Remove extra crumb from the surface. Apply syrup on both the surface of cake.

III For Icing:

1. Take 100gm cream and add icing sugar to it and beat it over ice bath till strong peak is seen. To check this stage, take the cream in a spoon, drop the cream if it doesn't fall, i.e., it forms the peak then it is considered that cream is made.

2. Put some icing mix on the cut and moistened cake surface.

3. Put the other layer on it and add rest of the cream over it. Level the icing and smoothen it.

4. Garnish the cake with pineapple and cherries.

Observations

S.No.	Ingredients	Weight of ingredients (gm)	%	Cost/kg (Rs.)	Cost (Rs.)
I.	**For Baking**				
2.	Maida	85			
3.	Vegetable oil	20 ml			
4.	Sugar	110			
5.	Egg	145			
6.	Baking Powder	2			
7.	Cake Gel	5			
8.	Salt	1			
9.	Water	40 ml			
10.	Cocoa Powder	15			
11.	Lemon and Chocolate Essence	Few drops			
II.	**For Icing**				
1.	Cream	100			
2.	Icing Sugar	50			
3.	Chocolate and Cherries	For garnishing			
III.	**For Syrup**				
1.	Water	100 ml			
2.	Sugar	15-20			
3.	Vanilla/Pineapple Essence	3-4 drops			

Notes:
1. Precaution should be taken while whipping the cream for icing as it may curdle.
2. Soft peak means that the peak can bend down whereas hard peak means peak gets permanently formed.
3. Beat the cream for icing very carefully in an ice as butter may get separated.
4. Do not open the oven in between the baking as it may lead to fall in the temperature.
5. Do not add too much of cocoa powder as it may compact the volume of cake.

Note your sensory observations below:

External	Internal
• Crust Colour	• Crumb Colour
• Volume Gain	• Symmetry of Form
• Aroma	• Sensory

Calculations

1. Over Head Cost (OHC) =

2. Total or production cost = Ingredient cost + O.H.C.

1. Baking/ Cooling losses
   = Wt. of dough–Wt. of final product

2. Baking-Cooling loss (%)

Practical No. 49

Aim: Preparation of Tutti Fruity Cake.

Procedure

1. Cream the margarine with sugar till a fluffy mix is formed.
2. Break the eggs in a bowl and beat them properly. Take 2-3 teaspoon of liquid egg to another bowl and add cake gel to it, mix it thoroughly to a stiff peak stage. To the rest of the egg add appropriate amount of both the essence and mix well.
3. Add the egg plus cake gel mixture to the creamy mixture and mix it thoroughly using the kitchen aid machine and then add the egg with essence, blend it well with the help of machine.
4. Sieve the maida and take 2 teaspoon full of sieved maida and add this to weighed amount of tutti fruity. To rest of the maida add baking powder and sieve it atleast thrice so that thorough mixing can be achieved.
5. Add this sieved maida mixture to the above creamy batter along with adding tutti fruity alternatively.
6. Mix it properly using cut and fold method by hand.
7. Pour this mixture to lined or greased tin and bake it at temperature of 190°C for 35 to 40 minutes.
8. Check the doneness of cake by inserting a toothpick. The toothpick should come out clean and be free from crumb particles.
9. After baking, take out the cake from the cake tin and allow it to cool.
10. After cooling, slice the cake and serve.

Observations

S.No.	Ingredients	Weight of ingredients (gm)	%	Cost/kg (Rs.)	Cost (Rs.)
1.	Maida	100			
2.	Margarine/ HVO	80			
3.	Sugar	100 to105			
4.	Egg	90 (~2 egg)			
5.	Baking Powder	2 to 3			
6.	Cake Gel	3.5			
7.	Tutti Fruity	25			
8.	Pineapple Essence	2-3 drops			
9.	Lemon Essence	4-6 drops			

Notes:

1. Level the cake batter by using milk (apply milk wash after pouring the cake batter in the tin). This will impart colour due to maillard browning and give glossy appearance to the cake top.
2. Tutti fruity is first coated with maida and is then added to the cake. If this is not done then because of high density it may settle down to the cake bottom and will not be evenly distributed in the cake.
3. Tutti fruity cake is a rich cake, rich cakes should be baked at lower temperature for a longer period of time compared to lean cake. This is because the batter of such cakes is very thick, so it takes time to cook. And if baking is done at higher temperature then the crust layer may be formed early which may lead to improper moisture removal from the cake center/core.

Note your sensory observations below:

External	Internal
• Crust Colour	• Crumb Colour
• Volume Gain	• Symmetry of Form
• Aroma	• Sensory

Calculations

1. Over Head Cost (OHC) =	1. Baking/ Cooling losses = Wt. of dough–Wt. of final product
2. Total or production cost = Ingredient cost + O.H.C.	2. Baking-Cooling loss (%)

Practical No. 50

Aim: Preparation of Marble Cake.

Procedure

1. Cream the margarine with sugar till a fluffy and creamy mixture is formed by using kitchen aid mixer.

2. Break the eggs in a bowl and beat them properly. Take 2-3 teaspoon of liquid egg to another bowl and add cake gel to it, mix it thoroughly to a stiff peak stage. To the rest of the egg add appropriate amount of both the essence and mix well.

3. Add the egg plus cake gel mixture to the creamy mixture and mix it thoroughly using the kitchen aid machine and then add the egg with essence, mix egg at slow speed of 2 followed by mixing at higher speed of 4 with the help of kitchen aid mixer.

4. Sieve the maida and baking powder atleast thrice so that thorough mixing can be achieved. Then divide it into two equal parts.

5. To one part add cocoa powder and then again sieve it thrice for proper mixing.

6. Divide the creamy mixture into two equal parts. To one part add maida and baking powder mixture and mix it well by using cut and fold method preferably by hand.

7. To the remaining part of creamy mixture, add maida, baking powder and cocoa powder mixture. Mix it well by using cut and fold method preferably by hand.

8. Pour this mixture into lined or greased tin in different patterns so as to give layers to the mix. After cake tin is ready tap it 1-2 times and bake it a temperature of 200 °C for 35 to 40 minutes.

9. Check the doneness of cake by inserting a toothpick. The toothpick should come out clean and be free from crumb particles.

10. After baking, take out the cake from the cake tin and allow it to cool.

11. After cooling, slice the cake and serve.

Observations

S.No.	Ingredients	Weight of ingredients (g)	Cost/kg (Rs.)	Cost (Rs.)
1.	Maida	95		
2.	Margarine/ HVO	80		
3.	Sugar	105		
4.	Egg	90 (~2 egg)		
5.	Baking Powder	2		
6.	Cake Gel	3.5		
7.	Cocoa Powder	7.5		
8.	Vanilla Essence	Few drops		
9.	Lemon Essence	Few drops		
10.	Chocolate Essence	Few drops		

Notes:

1. Essence should be added carefully as if it is added in excess then it may result in bitterness.
2. While baking the cake tin should be covered with the lid so that cracks or bulging appearance in the center of the cake can be avoided.
3. Layering should be done in such a way that different layers can be observed clearly in the cake slice. And cake should have both the flavours without intermixing.
4. If curdling is observed in the creamy batter, then either churn it at a speed of 4 then at 6 or add 1 teaspoon of maida to it.

Note your sensory observations below:

External	Internal
• Crust Colour	• Crumb Colour
• Volume Gain	• Symmetry of form
• Aroma	• Sensory

Calculations

1. Over Head Cost (OHC) =	1. Baking/ Cooling losses = Wt. of dough- Wt. of final product
2. Total or production cost = Ingredient cost + O.H.C.	2. Baking-Cooling loss (%)

Part-II
Analytical Bakery

EXPERIMENTS

Experiment-1

Aim: Determination of moisture content in food products by hot air oven- drying method.

Principle: Water plays an important role in bakery production. Because during processing water is the main factor in controlling in rheological properties of the dough. Moreover, its presence in the baked products influence on their palatability, freshness and keeping quality. Hence, it is necessary to determine the moisture level. This procedure is based on the determination of the loss in mass on drying of food product under specified condition. This loss of amount gives a measure of moisture present in the material.

Apparatus

1 Moisture dish - made of porcelain, silica, glass or aluminium.
2 Electric oven maintained at 105 ± 1°C.
3 Desiccator

Procedure

1. Weigh accurately about 5g of sample in the moisture dish which is previously air dried in the oven and weighed.
   2. Place the dish in the oven maintained at 105 ± 1°C for 4 hours. For flour sample, take 2 gm of well mixed sample into dry silica dishes. Dry dishes in oven at approximately 130°C for 1 hour.
3. Cool the dish in the desiccator and weigh.
4. Repeat the process of drying, cooling and weighing at 30 minutes. intervals until the difference between the two consecutive weighing is less than 1mg.
5. Record the lowest weight and calculate the moisture as per the below formula:

Observations

S.No.		Sample A	Sample B
1.	Weight of moisture dish.		
2.	Flour weight		
3.	Dish + dry flour weight		
4.	Dish + flour		
5.	Evaporated Moisture		

Calculations

Moisture, % by mass=100 × $(W_1-W_2)/W$
Where,
W_1= Weight in g, of the dish with the material before drying,
W_2= Weight, in g, of the dish with the material after drying, and
W = Weight, in g of the empty dish.

Results and Inference

Calculate the mean of the values obtained in the 2 determination and result is expressed in percentage (m/m). PFA limit-14 per cent max for wheat flour.

Precautions

1. Use a calibrated analytical balance capable of weighing to an accuracy of 0.001g.

   2.Coarse samples should be grinded rapidly and uniformly, without appreciable development of heat.
3. Use dish having an effective surface area enabling the test portion to be distributed so as to give a mass per unit area of not more than $0.3g/cm^2$.

Note: This method is not suitable for determination of moisture in foods like milk products or mineral mixture.

Sample dishes in moisture oven

Ref: SP: lS (Part I)-1980.

Experiment-2

Aim: Determination of ash content in flour sample.

Principle: Ash content is determined by high temperature incineration in muffle furnace. When a sample is incinerated in an ash oven, the high temperature drives out the moisture and burns away all the organic materials (starch, protein, and oil), leaving only the ash. The residue (ash) is composed of the non-combustible, inorganic minerals that are concentrated in the bran layer. Ash content results for wheat or flour ash are expressed as a percentage of the initial sample weight.

Importance of ash content in the flour sample: The ash content in wheat and flour has significance for milling and bakery industry. Baker needs to know the overall mineral content of the wheat to achieve desired or specified ash levels in flour. Since ash is primarily concentrated in the bran, ash content in flour is an indication of the yield that can be expected during milling. Ash content also indicates milling performance by giving the amount of bran contamination in flour. Ash in flour can affect colour, imparting a darker colour to finished products. Some products requiring white flour require low ash flour while other products, such as whole wheat flour, have high ash content.

The mineral constituents of wheat are not uniformly distributed throughout the kernel. Generally the bran contains about 5-8 per cent whereas endosperm contains 0.3 – 0.4 per cent, i.e., the bran contains large quantity. So the quantity of ash indicates the degree of endosperm separation from the bran during milling, i.e., the grade of the flour.

Apparatus

1. Flat bottom dish - of stainless steel, porcelain, silica or platinum.
2. Muffle furnace maintained at 550 ± 10°C.
3. Desiccators.
4. Tongs.

Procedure

1. Weigh accurately about 3-5 g of the sample in the dish, previously dried in an air-oven and weighed. If the sample is biscuit or any other product then first grind and take homogeneous sample.
2. Heat the dish gently on a flame at first and then strongly in a muffle furnace at 550±20°C till grey ash results.
3. Cool the dish in desiccators and weigh. Heat the dish again at 550±20°C for 30 minutes. Cool the dish in a desiccator and weigh. Repeat this process of heating for 30 minutes, cooling and weighing until the

difference between two successive weighings is less than one milligram.

4. Record the lowest weight.

Observations

S.No.		Sample A	Sample B
1.	Empty Dish.		
2.	Flour weight		
3.	Empty Dish + Flour weight		
4.	Empty Dish + Ash Weight		
5.	Ash content in flour		

Calculations

Total ash, percent by weight = $100 \ (W_2 - W) \ /W_1 - W$

Where,

W = weight in g of the empty dish,

W_1 = weight in g of the dish with the ash, and

W_2 = weight in g of the dish with the material taken for the test.

1. Ash content = (Empty dish + Ash Weight) – Empty Dish

2. $\dfrac{\text{Ash Weight} \ \times \ 100}{\text{Flour Weight}} = \%$

NOTE: Preserve the dish containing the ash for the acid insoluble ash determination.

Results and Inference: Wheat flour ash is usually expressed on a common moisture basis. For example, on the basis of 14 per cent for wheat flour. FSSA limit : Max-1 per cent (Maida)

Precautions

1. Ash the sample properly till grey ash.

2. Weigh till the difference between two successive weighing is less than one mg.

3. Don't take out the crucibles from muffle furnace with bare hands as it is very hot, use tongs.

Crucibles in the muffle furnace

Ref.: SP: lS (Part I)-1980.

Experiment-3

Aim: Determination of acid insoluble ash (AIA) content in flour sample.

Apparatus:

1. Flat-Bottom Dish – of stainless steel, porcelain, silica or platinum.
2. Muffle Furnace – maintained at 550 ± 10°C.
3. Desiccator
4. Tongs

   Reagent: Dilute hydrochloric acid-5N, prepared from conc. hydrochloric acid.

Procedure

1. To the ash contained in the dish (see Note under 2.4), add 25 ml of dilute hydrochloric acid, cover with a watch-glass and heat on a water-bath for 10 minutes.
2. Allow to cool and filter the contents of the dish through a whatman filter paper No. 42 or its equivalent.
3. Wash the filter paper with water until the washings are free from the acid and return them to the dish. Keep it in an oven maintained at 100 ± 2°C for about 3 hours. Ignite in a muffle furnace at 550 ± 10°C for one hour.
4. Cool the dish in a desiccator and weigh. Heat the dish again at 550 ± 10°C for 30 minutes, cool in a desiccator and weigh.
5. Repeat this process of heating for 30 minutes, cooling and weighing until the difference between two successive weighing is less than one milligram.
6. Record the lowest weight.

Calculations

$$\text{Total ash, percent by weight} = 100 \ \frac{(W_2 - W)}{W_1 - W}$$

Where,

W_1 = weight in g of the dish with the acid insoluble ash,

W = weight in g of the empty dish, and

W_2 = weight in g of the dish with the material taken for the test (refer 2.4).

   Results and Inference: AIA is usually expressed on a common moisture basis. For example on the basis of 14 percent for wheat flour. FSSA limit : Max-0.1 per cent (Maida).

Precautions

1. Carefully handle the HCl acid.

Ref.: BIS SP: lS (Part I)-1980

Experiment-4

Aim: Determination of protein content in food products by kjeldahl method.

Principle: In the kjeldahl Procedure, proteins and other organic food components in a sample are digested with sulphuric acid in the presence of catalysts. The total organic nitrogen is converted to ammonium sulphate. The digest is neutralized with alkali and distilled into a boric acid solution. The borate anions formed are titrated with standardized acid, which is converted to nitrogen in the sample. The result of the analysis represents the crude protein content of the food since nitrogen also comes from non-protein components (note that the kjeldahl method also measures nitrogen in any ammonia and ammonium sulphate).

Procedure and Reactions

Sample Preparation: Solid foods are ground to pass a 20-mesh screen. Samples for analysis should be homogeneous. No other special preparations are required.

1. Digestion:

Digestion with sulphuric acid, with the addition of powdered potassium permanganate to complete oxidation and conversion of nitrogen to ammonium sulphate.

Place sample (accurately weighed) in a Kjeldahl ask. Add acid and catalyst; digest until clear to get complete breakdown of all organic matter. Nonvolatile ammonium sulfate is formed from the reaction of nitrogen and sulfuric acid.

$$\text{Protein} \xrightarrow[\text{Heat, Catalyst}]{\text{Sulphuric acid}} (NH_4)_2SO_2 \qquad \dots [1]$$

During digestion, protein nitrogen is liberated to form ammonium ions; sulphuric acid oxidizes organic matter and combines with ammonium formed; carbon and hydrogen elements are converted to carbon dioxide and water.

2. Neutralization and Distillation

Neutralization of the diluted digest, followed by distillation into a known volume of standard acid, which contains potassium iodide and iodate. The digest is diluted with water. Alkali-containing sodium thiosulfate is added to neutralize the sulphuric acid. The ammonia formed is distilled into a boric acid solution containing the indicators methylene blue and methyl red (AOAC Method 991.20).

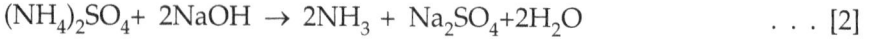

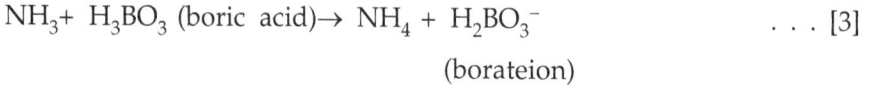

$$(NH_4)_2SO_4 + 2NaOH \rightarrow 2NH_3 + Na_2SO_4 + 2H_2O \qquad \ldots [2]$$

$$NH_3 + H_3BO_3 \text{ (boric acid)} \rightarrow NH_4 + H_2BO_3^- \qquad \ldots [3]$$

$$\text{(borateion)}$$

3. Titration

Titration of the liberated iodine with standard sodium thiosulphate.

Borate anion (proportional to the amount of nitrogen) is titrated with standardized HCl.

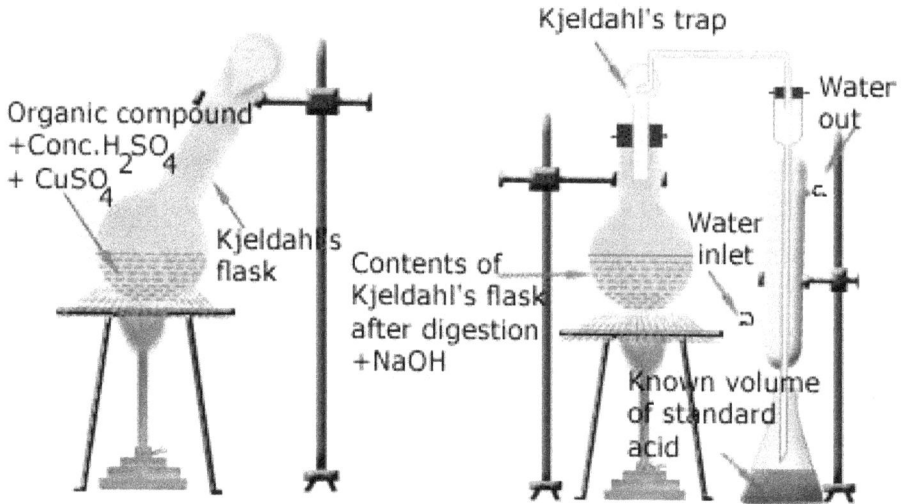

$$H_2BO_3^- + H^+ \rightarrow H_3BO_3 \qquad \ldots [4]$$

Fig. 4.1. Distillation Assembly

4. Calculations

$$\text{Moles of HCl} = \text{moles of } NH_3$$
$$= \text{moles of N in the sample} \qquad \ldots [5]$$

A reagent blank should be run to subtract reagent nitrogen from the sample nitrogen.

$$\% N = N \text{ HCl} \times \frac{\text{Corrected acid volume}}{\text{g of sample}} \times \frac{14 \text{ g N}}{\text{mol}} \times 100 \qquad \ldots [6]$$

where:

$$N \text{ HCl} = \text{normality of HCl,}$$
$$\text{in mol}/1000\text{ml}$$

Corrected acid vol. = (ml std. acid for sample) – (ml std. acid for blank)

14 = atomic weight of nitrogen

A factor is used to convert percent N to percent crude protein. Most proteins contain 16 per cent N, so the conversion factor is 6.25 (100/16= 6.25).

per cent N/0.16 = per cent protein   . . . [7]

or

per cent N × 6.25 = per cent protein

Conversion factors for various foods are given in Table below:

Nitrogen to Protein Conversion Factors for Various Foods

	Percent N in Protein	Factor
Egg or meat	16.0	6.25
Milk	15.7	6.38
Wheat	18.76	5.33
Corn	17.70	5.65
Oat	18.66	5.36
Soybean	18.12	5.52
Rice	19.34	5.17

Results and Inference: Report the result and draw inference of sample with reference to the standard.

Precautions

1. Handle acid safely: use acid resistant fume hood. Always add acid to water unless otherwise directed in method. Wear face shield and heavy gloves to protect against splashes. If acids are spilled on skin, immediately wash with large amounts of water.

2. Sulphuric acid and sodium hydroxide can burn skin, eyes and respiratory tract severely. Wear heavy rubber gloves and face shield to protect against concentrated acid or alkali. Use effective fume removal device to protect against acid fumes or alkali dusts or vapours. Always add concentrated sulphuric acid or sodium hydroxide pellets to water, not vice versa. Concentrated sodium hydroxide can quickly and easily cause blindness. If splashed on skin or in eyes, flush with copious amounts of water and seek medical attention.

3. Keep baking soda and vinegar handy in case of chemical spills.

4. The sulphur oxide fumes produced during digestion are hazardous to breathe. Do not inhale.

5. Digests must be cool before dilution water is added to avoid a violent reaction during which the acid can shoot out of the flask. Likewise,

the diluted digest must be cool before sodium hydroxide is added to avoid a similarly violent reaction.

Advantages of Method

1. Applicable to all types of foods.
2. Inexpensive (if not using an automated system)
3. Accurate; an ofcial method for crude protein content
4. Has been modied (micro Kjeldahl method) to measure microgram quantities of proteins.

Disadvantages of Method

1. Measures total organic nitrogen, not just protein nitrogen.
2. Time consuming (at least 2h to complete).
3. Poorer precision than the biuret method.
4. Corrosive reagent.

Ref.: Food Analysis, Neilson, Fourth Edition p. 136.

Experiment-5

Aim: Determination of falling number of wheat flour sample.

Principle: The falling number test provides an index of α-amylase in a flour or ground-wheat sample. The Procedure is based on the reduction in viscosity of starch paste caused by the action of á amylase. The method is based on the unique ability of alpha-amylase to liquefy a starch suspension. Gelatinization strength is measured by falling number as "time in seconds" required stirring and allowing the stirrer to fall a measured distance through hot aqueous flour gel undergoing liquefaction.

Requirement: Falling number instrument, viscometric tube.

Procedure

1. The distilled water in bath is brought to boil.
2. Weigh 7 gm of flour, transfer it to viscometric tube, and add 25 ml of distilled water, rubber the tube and shake vigorously for obtaining a uniform suspension.
3. Remove stopper and push down flour adhering to sides with the viscometer stirrer.
4. Place the tube with stirrer in the boiling water bath. Start the timer.
5. After 5 seconds, automatic stirring starts at the rate of 2 stirs/seconds for 60 seconds. After a total of 60 seconds, stirring automatically stops releasing the stirrer at its uppermost position and allows falling by its weight at a fixed distance.
6. Time is recorded in seconds.

Result and Inference

Note the time in seconds and draw the inference.

Importance: Starch provides the supporting structure of bakery product, too much enzyme activity results in sticky dough during processing and poor texture in the finished product. The level of enzyme activity measured by the falling number test affects product quality. Yeast in bread dough, requires sugars to develop properly and therefore needs some level of enzyme activity in the dough. Too much enzyme activity however, means that too much sugar and too little starch are present. If the falling number is too high, enzymes can be added to the flour in various ways to compensate. If the falling number is too low, enzymes cannot be removed from the flour or wheat, which results in a serious problem that makes the flour unusable.

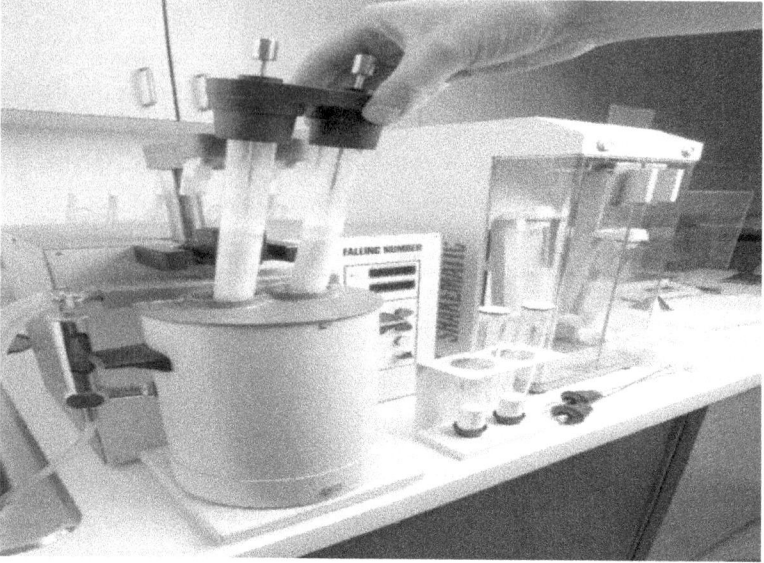

Ref: AACC Approved Method 56-81B- wheat flour notes Adapted from Method 56-81B, Approved Methods of the American Association of Cereal Chemists, 10th Edition. 2000, St. Paul, MN.

Experiment-6

Aim: Determination of thousand kernel weight of wheat grain sample.

Requirement: Thousand kernel weight instrument, wheat grain sample.

Procedure

1. Prepare a 500-gram sample of wheat by removing all dockage, shrunken and broken kernels, and other foreign material.
2. Divide the sample several times using a mechanical divider until you have approximately 50 grams.
3. Count 1,000 kernels using a mechanical counter and weigh.

   Results: Note the weight and record.

Importance: Thousand kernel weight (TKW), measures the weight of the wheat kernel. It is used by flour millers as a complement to test weight to describe wheat kernel composition and potential flour extraction. So, wheat with a higher TKW can be expected to have a greater potential flour extraction.

Thousand kernal weight instrument

Experiment- 7

Aim: Determination of water absorption power and gluten quantity in the wheat flour sample.

Requirement: Beaker-500ml, measuring cylinder, muslin cloth, dish, air oven.

Principle: To separate gluten from other constituents, the wheat flour is mixed with water. The native proteins of flour interact to form a chewing gum type of wet mass, which is called wet gluten. The wet gluten can be washed out using potable water or using automatic gluten washer. The wet gluten is dried to form a free flowing light coloured powder. Depending upon the wheat variety, wide variation in the quality of extracted gluten are observed.

The procedure is applicable to whole wheat and refined wheat flour. The dough developed by mixing wheat flour with water possesses the viscoelastic characteristics vital for dough handling and final product quality. The viscoelastic nature of dough is attributed to gluten proteins namely gliadins and glutenins. The gliadins impart extensibility to dough, whereas glutenin is held responsible for strength and elastic character of gluten and dough.

Procedure

1. Weigh 50g sample of wheat flour and put it into a bowl.

   2.Add water from burette to know the percentage of water absorption and knead it.

3. Make a soft dough ball. Further knead it to convert into a smooth ball. Immerse it in sufficient water for 1 hour to hydrate starch and form gluten network.

4. Wash this dough ball with intermittent kneading to remove starch with muslin cloth. Taking care to see that no gluten is lost while discarding the wash water. The washings are continued till the Gluten is free from starch. Check the development of blue-black colour for the presence of starch using iodine solution.

5. Stretch the gluten to know its quality. Squeeze the gluten ball as much as possible and make it uniform.

6. After washing the wet gluten is placed on a piece of tared and previously weighed dish or glass petriplate.

7. Weigh the dish with wet gluten to find the per cent wet gluten.

8. Knead it to form a smooth ball and keep in a preheated oven for 15 minutes at 220-230°C.

9. Dry it after pricking with a pin randomly so that there is an easy transfer of heat and moisture from the centre to periphery.

10. Reduce the temperature to 100°C and keep from proper dryting.
11. Take out the petriplate and cool it in the dessicator. Weigh to find the per cent dry gluten.
12. Calculate the per cent water absorption.

Calculations

Wet gluten (%) = $\dfrac{A \times 100}{C}$

Dry gluten (%) = $\dfrac{B \times 100}{C}$

where,
A = wt. of wet gluten (g),
B = wt. of dry gluten (g), and
C = wt. of flour (g).

$$\text{per cent water absorption} = \frac{\text{volume of water used (ml)}}{\text{Weight of flour (g)}} \times 100$$

Results and Inference: Calculate the weight of gluten and infere the result with the standard.

Precautions

1. Care should be taken that no dough particle is washed off.
2. Dough should be washed till the gluten is free from starch and is confirmed by using iodine solution.

Experiment-8

Aim: Estimation of volume and specific gravity of flour and cornflour.

Introduction : The specific gravity of flour, corn flour and sugar is indirectly proportionate to the volume of flour, corn flour and sugar.

Apparatus: Beaker, Weighing scale.

Materials: Water, flour, Corn Flour, (Sample)

Procedure

1. Keep ready the flour and corn flour
2. Weigh the empty beaker first.
3. Then fill the beaker with flour and corn flour are by one and weigh it properly.
4. Then empty the beaker, clean and dry it.
5. Now fill it with water to the same level of flour and corn flour and weigh again.

Observations

1. Weight of dry empty beaker   :   gm
2. Weight of beaker filled with flour   :   gm
3. Weight of beakers filled with corn flour   :   gm
4. Weight of beaker filled with water   :   ml

Calculations: Calculate the density and specific density of flour as below:

1. Weight of flour = weight of beaker filled with flour – weight of empty beaker
2. Weight of water = Weight of beaker filled with water – Weight of empty beaker.

$$\text{Density} = \frac{\text{Weight}}{\text{Volume}} = \text{cubic cm.}$$

$$\text{Sp. Gravity} = \frac{\text{Weight of flour}}{\text{Weight of same volume of water}} =$$

Experiment-9

Aim: Qualitative estimation of flour refineness by peckar colour test.

Introduction: The colour of wheat flour has significance importance, as it effects on the ultimate crumb colour of the baked leaf and it is index to flour grade, effectiveness of bleach treatment and degree of granularity.

Apparatus : Glass plate, Beaker, Steel slick.

Materials: Flour, Water, Pyrocatechol solution (1 per cent).

Procedure

1. Take glass plate.
2. Apply little water on both the edges of glass slide and make it wet.
3. Place a heap of one sample of flour at the left end of glass plate at least (1/2") at one end.
4. Next place a heap of another sample of flour at the right hand of the glass plate at least (1/2") at the end of plate.
5. Case should be taken properly that the sample of both end of slide should not mix with each other.
6. The samples are gently submerged at an angle in fresh clean water until air bubbles cease to rise.
7. Quickly withdraw and transfer into a warm place.
8. Note the change and relative intensity of colour and compose the flour again.

Observations:

Sample No.	Type of flour	Colour when dry	Colour after being wet with water (and removed) from oven.	Colour after being wet with pyrocatechol (and removed from oven)
1.	Sample -1			
2.	Sample -2			

Experiment-10

Aim: Determination of sedimentation value of the flour.

Apparatus : 100 cc glass stoppered measuring cylinder stopwatch, pipette (25cc) glass rod.

Materials: Flour, Distilled Water, Lactic acid solution.

Procedure

1. Add 50 cc distilled water to a 100 cc glass stoppered graduated cylinder.

2. Place exactly weighed 4 gm of the flour under test into the glass cylinder.

3. Shake the mix. With glass rod for 30 sec. and allow standing for 5 minutes.

4. Add 25cc of lactic acid solution by means of a pipette, mix the contents of the cylinder by inverting and returning it to an upright position 10 times.

5. Place the cylinder in an upright position in front of light and start time.

6. After an interval of exactly 5 min read the volume of the solid phase of the material in the cylinder.

7. This volume in ml is the sedimentation value of the flour.

No.	Type of flour	Weight	Sedimentation Value
1	A	4 gm	.....cm

Note : The sedimentation value can give an estimation on quality and quantity of flour. The sedimentation value can be interpreted as below:

Sedimentation Value	Remarks
20 – 55	Good quality of flour
<20	Soft flour (low protein)
>55	High protein (Superior bread baking quality)

Experiment-11

Aim: Determination of yeast quality by its dough rising capacity.

Introduction: Yeast is the principal leavening agent in the fermented products like bread, bun, toast etc. The structure of such product is depends on the speed and degree of fermentation. Hence, if the quality of yeast is proper the fermentation will be proper and finally the products quality will also be proper.

Apparatus : Beaker, Ruling scale.

Materials: Flour, water, yeast and sugar.

Procedure

1. Prepare dough having similar consistency to bread from 100 gm flour, 1 to 1.5 g sugar and 2 gm fresh yeast and approximately 55 ml water.
2. Place it in a beaker in such a way that the upper surface of dough becomes even, and no air entrapped within dough.
3. The height of beaker should at least 2½ to 3 times the dough height.
4. Measure the height of the dough placed in a beaker.
5. Keep it aside at normal room temperature and measure the height again after 1 hour.
6. The difference between two heights is nothing but dough rising capacity of the yeast.

Observations

1. Height of dough at beginning : ...... **cm**
2. Height of dough after 1 hr. : ......**cm**
   (i) Increase in dough height = Height of dough after 1 hr. – Height of dough at beginning = ...
   (ii) Dough rising capacity of the yeast (per cent)

$$= \frac{\text{Increase in dough height}}{\text{Height of dough at beginning}} \times 100$$

Experiment-12

Aim: Determination of starch in cereal grains by acid hydrolysis method.

Principle : Cereal grains usually contain as high as 70 per cent of starch. Starch is a chief calorific constituent of food. The manufacture of malt liquors, alcohol, distilled liquors and vinegar, so far as these products are made from cereal grains, the quality is dependent on the starch content of the raw material. The value of the raw materials used, other nutritional parameters being equal, is proportional to the starch content.

Upon treatment with acid, the components of starch (amylase and amyl pectin) are hydrolyzed progressively via dextrin and maltose to the final product (glucose). In order to have complete hydrolysis, the sample needs to be defatted and the acid hydrolysis process is carried out by refluxing for at least 2.5 h.

Requirements

(a) Regents

1. Diethyl Ether, AR grade
2. Ethyl Alcohol (10 per cent, v/v).
3. *Dilute Hydrochloric Acid (2.5 per cent):* Prepared by mixing 20ml of concentrated hydrochloric acid (specific gravity 1.16) and 200ml of water.
4. Sodium Carbonate Solution (20 per cent m/v).
5. Stock Solution of Dextrose- Weigh accurately 10g of anhydrous dextrose into a one-litre graduate flask and dissolve it in water. Add to this solution 2.5 g of benzoic acid, shake to dissolve benzoic acid and make up the volume to the mark with water. (This solution should not be used after 48 hours).
6. *Standard Dextrose Solution:* Dilute a known aliquot of the stock solution of dextrose (with water containing 0.25 per cent, m/v of benzoic acid to such a concentration that more than 15 ml but less than 50 ml of it will be required to reduce all the copper in the Fehling's solution taken for titration. Note the concentration of anhydrous dextrose in this solution as mg/100ml (Prepare this solution fresh everyday).
7. *Methylene Blud Indicator Solution:* Dissolve 0.2 g of methylene blue in water and dilute to 100 ml.
8. *Fehling's Solution (Soxhlet modification):* Prepare by mixing immediately before use, equal volume of solution A and solution B.
8a. *Solution A:* Dissolve 34.639g of copper sulphate ($C_uSO_45H_2O$) in water, add 0.5 ml of concentrated sulphuric acid of sp gr 1.84 and dilute to 500 ml in a graduated flask. Filter the solution through prepared asbestos.

8b. *Solution B:* Dissolve 173 g of potassium sodium tartrate (K-Na-$C_4H_4O_6.4H_2O$) and 50g of sodium hydroxide in distilled water, dilute to 500 ml in a graduated flask and allow the solution to stand for 2 days. Filter this solution through prepared asbestos (washed).

9. *Carrez-I (Zinc Acetate Solution):* Dissolve 21.9 g of zinc acetate [Zn $(C_2H_3O_2)_2.2H_2O$] and 3 ml of glacial acetic acid in water. Dilute to 100 ml.

10. *Carrex-II (Potassium Ferrocyanide Solution):* Dissolve 10.6 g postassium ferrocyanide in distilled water and make the volume to 100 ml.

Procedure

1. Standardization of Fehling's Solution: Pour the standard dextrose solution into a 50 ml burette. Find the titre, i.e., the volume of the standard dextrose solution required to reduce all the copper in 10 ml of Fehling's solution corresponding to the concentration of the standard dextrose solution. Pipette 10 ml of Fehling's solution into a 300 ml conical flask and add from the burette, standard dextrose solution required for reduction of all the copper, so that not more than 1 ml will be required later to complete the titration. Heat the flask containing the mixture over wire gauze. Gently boil the contents of the flask for 2 minutes. At the end of 2 minutes of boiling, add 1 ml. Of methylene blue indicator solution without interrupting boiling. While the contents of the flask continue to boil, begin to add standard dextrose solution (one or two drops at a time) from the burette till the blue colour of the indicator just disappears.

Note: The titration should be completed within 1 minute, so that the contents of the flask boil altogether for 3 minutes without interruption. Note the titre. Multiply the titre (obtained by direct titration) by the number of milligrams of anhydrous dextrose in 1 ml of the standard dextrose solution to obtain the dextrose factor.

2. Preparation of the Solution: Weigh 5g of ground material add 220 ml of 2.5 per cent dilute HCl. Attach to reflux condenser and reflux for 2½ hours. Transfer the contents to 500 ml volumetric flask. Add 5 ml Carrez-I and 5 ml of Carrez-II. Make up the volume to 500 ml and filter using Whatman No. 40 filter paper. Discard first few drops. Take 50 ml of the filtrate in 250 ml volumetric flask, add about 150 ml water and neutralize. Make up the volume to 250 ml. complete titration in the same manner as for the standard.

Calculations

$$\text{Starch, \% by mass} = \frac{\text{Std titre Std conc} \left(\frac{mg}{ml}\right) \times \text{volume made} \times 0.93}{\text{Sample titre} \times \text{sample wt (g)} \times 10}$$

Results and Inference: The mean of the results of two determinations should be reported. Cereal grains intact with bran have lower starch content as compared to refined and processed grains.

Precautions

1. The sample should be ground to powder so as to pass through 40 mesh size (<0.5mm) before analysis to aid in complete hydrolysis.

   2.During hydrolysis, avoid frothing by regulating the heat intensity so that contents just keep simmering.

3. Ensure neutralization of acid with alkali before titration.

4. Titration should be performed in such way that it completes within 1 minutes. Thereby minimizing the oxidation due to air.

Ref.: IGNOU study material.

Experiment 13

Aim: Determination of the salt content in finished bakery products.

Principle: This method determines the salt concentration of a solution by titration with silver nitrate. As the silver nitrate solution is slowly added, a precipitate of silver chloride forms.

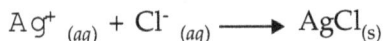

$$Ag^+_{(aq)} + Cl^-_{(aq)} \longrightarrow AgCl_{(s)}$$

The end point of the titration occurs when all the chloride ions are precipitated. Then additional chloride ions react with the chromate ions of the indicator potassium chromate, to form a red brown precipitate of silver chromate.

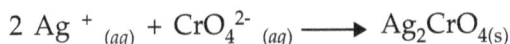

$$2\ Ag^+_{(aq)} + CrO_4^{2-}_{(aq)} \longrightarrow Ag_2CrO_{4(s)}$$

Requirements

(a) Reagents

1. Silver nitrate solution (0.1N) – Weigh about 17.5 g of $AgNO_3$ (Merck or Loba Chemie -AR grade) and make up the volume to 1 litre with distilled water in a volumetric flask. Store the solution in a brown bottle.

2. 5 per cent Potassium chromate indicator solution in water.

   Standardisation of Silver Nitrate

   Weigh about 5.844 g of sodium chloride previously dried and make up the volume to 1 litre with distilled water. Take 50 ml of this and add 1 ml Potassium chromate indicator solution and titrate against $AgNO_3$ until the first perceptible pale red brown colour appears.

   (1 ml of 0.1 N Silver Nitrate is equivalent to 5.845/1000 g of sodium chloride).

   Calculation for Standardization of 0.1N Silver Nitrate:

   Normality of $AgNO_3$ = Volume of NaCl × Normality of NaCl (0.1 N)/Titre Value.

(b) Glasswares (Borosil grade): 50ml pipettes, 250 ml conical flask, burette, measuring cylinder and beakers.

(c) Sample preparation

1. Take 2-3g of powdered biscuit samples in a beaker. Add around 100 to 150 ml of distilled water to it and soak it for half-an-hour to one hour with intermittent stirring. Filter the liquid into a 200ml volumetric flask using a Whatman filter paper 541. Wash till the filtrate is free from chloride. (No white precipitate should be formed when the filtrate is reacted with silver nitrate solution).

2. Make up the volume to 200ml with distilled water.

3. Take an aliquot (50ml) of the sample extract into a 250ml conical flask and add 1 ml of Potassium chromate indicator solution.

4. Titrate against 0.1N AgNO$_3$ solution. The end point of the titration is identified by the first perceptible pale red brown colour appears. Repeat the titrations to get concordant value.

5. Carry out a blank titration taking 50ml of distilled water.

Calculations

$$\% \ NaCl = \frac{(\text{Titre value} - \text{Blank}) \times 5.844 \times 200 \times 100 \times \text{Normality of AgNO}_3)}{\text{Weight of sample} \times 50 \ (\text{aliquot}) \times 1000 \times 0.1}$$

Precautions

1. Avoid contact with Silver nitrate and wash immediately with water if AgNO$_3$ gets on skin or clothing.

2. White lumps of AgCl forms when titrating high concentration of salt. This should not be taken for the endpoint of the titration.

3. The burette should be rinsed out with distilled water immediately after the titration is completed.

Ref.: Handbook of Analysis and quality control for Fruit and Vegetable products, Second edition - S.Ranganna and AOAC.

Experiment-14

Aim: Determination of specific gravity of oils and fats.

Principle: The specific gravity bottle method is a gravimetric method in which the weight of sample is divided with the weight of water of same volume at same temperature. This method is more accurate and gives quick result. Specific gravity is usually determined with a specific gravity bottle or pyknometer.

The specific gravity of sample is calculated w.r.t the reference (water) which is 1 (or 1000 in british brewing) if and only if the reference and sample temperatures are the same.

Requirements

1. Specific gravity bottle or pyknometer- with well fitting ground glass joints.
2. Water bath - maintained at 30.0 ± 0.2 ° C, or 95.0 ± 0.2 °C as required.
3. Calibrated thermometer of a suitable range with 0.1 or 0.2°C subdivisions.

Procedure

1. **Calibration of pyknometer**: To calibrate, clean and dry the bottle or pyknometer thoroughly, weigh and then fill with recently boiled and cooled water at about 25°C after removing the cap of the side arm. Fill the over flowing by holding the bottle or pyknometer on its side in such a manner as to preen the entrapment of air bubbles. Insert the stopper and immerse in a water bath at the desired test temperature ±0.2°C.

   2.Keep the entire bulb completely covered with water and hold at that temperature for 30 minutes. Carefully remove any water, which has exuded from the capillary opening. Remove from the bath, wipe completely dry, replace the cap, cool to room temperature and weigh. Calculate the weight of water. This is a constant for the bottle or pyknometer, but should be checked periodically.

3. Melt the sample, if necessary, and filter through a filter paper to remove any impurities and the last traces of moisture, make sure that the sample is completely dry. Cool the sample to 30°C or warm to the desired test temperature. Fill the bottle with the oil previously cooled to about 25 °C or the melted fat to overflowing, holding the bottle on its side in such a manner as to prevent the entrapment of air bubbles after removing the cap of the side arm. Insert the stopper, immerse in the water bath at 30.0 ± 0. °C and hold for 30 minutes. Carefully wipe off any oil, which has come through the capillary

opening. Remove the bottle from the bath, clean and dry it thoroughly. Replace the cap of the side arm, cool to room temperature and weigh.

Results and Inference: Record the weight.

Precautions

1. Bottle should be dry.
2. Temperature should be maintained carefully.

Experiment-15

Aim: Determination of saponification value of oil sample.

Principle: The saponification value is related to the molecular weight of the fat, denotes the number of mg of potassium hydroxide which is required to saponify 1g of fat, *i.e.,* to neutralize the free fatty acids and the fatty acids combined as glycerides. The saponification equivalent is the amount of oil or fat saponified by 1 gram equivalent of potassium hydroxide, and is equal to 56.108 divided by the saponification value.

It gives information concerning the character of the fatty acids of the fat and in particular concerning the solubility of their soaps in water. The higher the saponification number of a fat free from moisture and unsaponifiable matter, the more soluble the soap that can be made from it. The information is of especial importance to soap makers.

Requirements

Apparatus: Conical flask-200ml, made of alkali resistant glass, provided with a reflux condenser with a ground joint.

Reagents:
(a) Hydrochloric acid - 0.5 N aqueous solution, accurately standardized.
(b) Potassium hydroxide - 0.5 N solution
(c) Phenolphthalein indicator - 0.1 per cent solution
(d) Neutralized (DN-DN spirit and n-propyl alcohol mixture (1 part spirit to dissolve KOH and 2 parts of n-propyl alcohol to dilute KOH 0.5N approx).

Procedure

1. About 1.5 to 2.0gm of the material is accurately weighed into 250ml flat bottom flask. 50ml of0.5N ethanolic potassium hydroxide is pipette into the sample flask and connected to the water reflux condenser.

2. The entire setup is refluxed on the electric mantle for 2 hours. Simultaneously 50ml of the ethanolic potassium hydroxide alone is pipette into another flatn bottom flask and is attached to another water reflux condenser electric mantle system for 2 hours refluxation.

3. To the hot test and blank solution 2-4 drops of phenolphthalein indicator is added and titrated against 0.5N standard hydrochloric acid from the 50ml volumetric burette. The end point is pink colour to colourless.

$$R\ COOH + KOH \longrightarrow R\ COOK + H_2O$$
$$KOH + HCl \longrightarrow KCl + H_2O$$

Calculations

Weight (in g) of oil or fat taken $\qquad$ = W

Volume (in ml) of hydrochloric acid used in test $\quad$ = V_1

Volume (in ml) of hydrochloric acid used in blank $\quad$ = V_2

Normality of hydrochloric acid $\qquad$ = N

$$\text{Saponification value} = \frac{56.1\ N\ (V_2\text{-}V_1)}{W}$$

Given the result correct 0.01 units

Results and Inference

Calculate the result and infere the result with standard value. Following table shows the saponification value of common fat or oil sample.

Saponification Numbers of Common Fats*

Fat or Oil	Saponification number
Rapeseed Oil	170 – 179
Menhaden Oil	190.6
Corn Oil	188 – 193
Olive Oil	185 – 196
Soy bean Oil	193
Cacao Butter	193.55
Linseed Oil	192 – 195
Cottonseed Oil	193 – 195
Lard	195.4
Mutton Tallow	192 - 195.5
Peanut Oil (arachis)	190 – 196
Horse Oil	195 – 197
Beef Tallow	193.2 – 200
Palm Oil	196 – 205
Butter	220 – 233
Palm Kernel Oil	242 – 250
Coconut Oil	246 – 260

Data from J. Lewkowitsch, *Chemical Technology and Analysis of Oils, Fats, and Waxes*, pp. 395-400.

Experiment-16

Aim: Determination of iodine value of oil and fat sample (WIJS METHOD).

Principle: Iodine value (IV) is a measure of the total number of double bonds present in fats and oils. It is generally expressed in terms of "number of grams of iodine that will react with the double bonds in 100 grams of fats or oils". High IV oil contains a greater number of double bonds than low IV oil. Edible oils with high iodine value are usually less stable and more susceptible to oxidation. The product to be investigated (oil, fatty acids) is treated with an iodine monochloride solution. After addition of the halogen has taken place, the excess of iodine monochloride is determined by titration with thiosulphate solution.

Apparatus: Conical flasks, 250ml with ground glass stoppers.

Reagents

  (a) Carbon tetrachloride/chloroform (AR Grade).

  (b) Potassium Iodide: 15 per cent aqueous solution, free from iodine and iodate

  (c) Sodium thiosulphate: 0.1N aqueous solution, accurately standardized.

  (d) Starch Indicator: stir 1gm, soluble starch in 3ml water and pour the mixture into 100ml boiling water. Keep the solution for boiling for 3minutes and cool.

  (e) Wij's solution (0.2 N ICI solution). This is commercially available. Commercially available ICl (39.44N) is first diluted to 500ml with glacial acetic acid (AR) in a standard volumetric flask. Shake this soln well, and 62.5ml of this soln. is further diluted to 500ml in a standard volumetric flask with glacial acetic acid. This reagent should be moisture free.

Procedure

  1. Around 0.2g in case of H.V.O. or 2.0g in the case of RCO is weighed into 250ml dried iodine flask (W).

  2. The contents are dissolved by adding 15ml carbon tetrachloride or chloroform. To that, solution 25ml wij's solution is added from the burette and allowed to stand in the dark at room temperature for 1hour after closing the flask with the stopper.

  3. Under similar condition a blank is prepared by adding 15ml carbon tetra chloride and 25ml of iodine mono chloride and kept inside dark at room temperature for 1hour in the mean while a volumetric 50ml burette is filled with 0.1 N thiosulphate solution and starch indicator

solution is kept ready for the titration.
4. After 1 hour standing one-by-one the flasks are removed from the dark and titrated against standard thio solution from the burette after the addition of 20ml of potassium iodide and about 50ml of water and starch indicator towards the end point of the titration.
5. The titration is continued till the disappearance of blue colour.

Calculations

Weight (in g) of sample taken $= W$

Volume (in ml.) of thiosulphate solution used in test $= V_1$

Volume (in ml.) of thiosulphate solution used in blank $= V_2$

Normality of thiosulphate solution $= N$

$$\text{Iodine Value} = \frac{12.69 \, N \, (V_2 - V_1)}{W}$$

Experiment 17

Aim: Determination of salt content in butter.

Principle: Salt is extracted in hot water and titrated with silver nitrate in presence of potassium chromate. The amount of silver nitrate (standard) used is calculated and expressed as percent sodium chloride.

Reagent

1. Potassium chromate - (K_2CrO_4) - 5 per cent solution
2. Standard Silver Nitrate solution - 0.1N.

Procedure

1. Weigh about 5g of sample into a 250ml conical flask. Carefully add 100ml of boiling water. Make it to stand 5-10min with occasional stirring.
2. Cool to about 50-55°C.
3. Add 2ml Potassium chromate solution. Mix by swirling. Add about 0.25gm of Calcium Carbonate and mix again.
4. Titrate with 0.1N $AgNO_3$ solution, while swirling continuously until a orange brown colour persists for 30 seconds.
5. Blank Test is carried out with all reagents in the same quantity. The maximum deviation between duplicate determinations should not exceed 0.02 per cent NaCl.

Calculations

$$NaCl\ (w/w) = \frac{12.69 \times N \times (V_1 - V_2)}{W}$$

where, N = Normality of Silver Nitrate

V_1 = Volume of Silver nitrate in sample titration

V_2 = Volume of Silver nitrate in blank titration

W = Weight in gm of the sample.

Experiment-18

Aim: Estimation of Peroxide Value of Oil Sample.

Principle: The Peroxide vale is determined by subjecting potassium iodide at room temperature to the oxidant effect of peroxides. The iodine thus liberated, is titrated with sodium thiosulphate.

Apparatus: Conical flask 250ml, with ground glass stoppers.

Reagents

1. Solvent-Mix 2 volumes of glacial acetic acid and 1 volume of chloroform.
2. Potassium Iodide-Saturated solution Dissolve 4 parts of pure potassium Iodide in 3 parts distilled water. Keep the solution in a brown bottle.
3. Sodium Thiosulphate-0.002N solution. Prepare this solution fresh from accurately standardized 0.1 N solution.
4. Starch Indicator-1 per cent solution, freshly prepared.

Procedure

1. About 2-3gms of the sample is accurately weighed into a dried 250ml iodine flask.

   2.To that 25ml of the solvent is added and mixed homogeneously.
3. 1 ml of the potassium Iodide solution is added and allowed to stand in the dark for one minute.
4. After one minute, 1ml of starch indicator is added and the solution is diluted with 35ml of distilled water and titrated against 0.002N solution of sodium thiosulphate from 10ml volumetric burette.
5. The end point is the disappearance of blue colour. A blank with 25ml chloroform acetic acid mixture is carried out.

Calculations

$$\text{Peroxide value} = \frac{1000 \times (V_2\text{-}V_1) \times N}{W}$$

Where, Weight (in g) of sample taken   = W
Volume (in ml) sodium thiosulphate used up by the sample = V_2
Volume (in ml) sodium thiosulphate used in blank   = V_1
Normality of sodium thiosulphate   = N

Note: Give results below 20 to the nearest 0.1, and above 20 to the nearest 0.5.

Ref.: SP: 18(Part XIII)-1984.

Experiment-19

Aim: Determination of crude fat in foods by soxhlet extraction method.

Principle: The crude fat content can be conveniently determined in foods by extracting the dried and ground material with petroleum ether or diethyl ether in Soxhlet extraction apparatus.

Extraction of the crude fat is carried out either with petroleum ether or diethyl ether in a Soxhlet unit followed by volatilization of the solvent after extraction and determination of the mass of the residue.

Requirements

(a) **Apparatus:** Soxhlet apparatus

(b) **Reagents:** Diethyl Ether – anhydrous or petroleum ether (bp. 60-80°C)

Procedure

1. Extract 2 g of the ground material in a continuous extraction apparatus with ether from 18 hours.

   2.Remove the ether by distillation, followed by blowing with a stream of air with the flask on a boiling water bath and dry in an oven at 110±1°C till the loss in mass between two successive weighing is less than 2 mg.

3. Shake the residue with 2 to 3ml of ether at room temperature, allow to settle and decant the ether. Repeat the extraction until no more of the residue dissolves.

4. Dry the flask again until the loss in mass between two successive weighing is less than 2 mg.

5. Record the final mass.

Calculations

$$\text{Crude fat, \% by mass} = \frac{(M_1 - M_2) \times 100}{2M}$$

where,

M_1 = mass, in g, of the soxhlet flask with the extracted fat.

M_2 = mass, in g, of the empty soxhlet flask, and

M = mass, in g, of the material taken for the test.

Results and Inference: Result of fat content should be reported to the nearest 0.1 per cent (m/m).

The method is only applicable to the food products having low moisture content and is suitable for the determination of free fat content (crude fat).

This method may also extract certain impurities soluble in ether.

Precautions

- The fat/oil obtained after drying should be clear and free from any particles. If the presence of particulate matter observed in the fat, the fat should be dissolved in petroleum ether again and filtered into other conical flask and dried.
- If the charring of fat is observed during drying, discard the fat and repeat the experiment.

Ref: IGNOU chemical analysis manual.

Experiment-20

Aim: Determination of pH of food products by using pH meter.

Principle: The pH value or hydrogen ion concentration is a measure of the acidity or alkalinity (basicity) of a solution. A neutral solution has a pH value of 7, an acid solution has pH value less than 7, and a basic solution has a pH value greater than 7. A change of one pH unit corresponds to a 10-fold change of hydrogen ion concentration of the solution.

It is expressed as: pH= −log [H+], where (H+) is a hydrogen ion concentration of solution in moles per litre. The pH value is determined by measurement of the electromotive force of a cell consisting of an indicator electrode (an electrode responsive to hydrogen ions viz., glass electrode) immersed in the test solution and a reference electrode (colomel electrode), contact between the test solution and the reference electrode is usually achieved by means of a liquid junction, which forms part of the reference electrode. The electromotive force measured with a pH meter, that is, a high impedance voltmeter calibrated in terms of pH.

Requirements

(a) **Apparatus:** Calibrated electrodes and potentiometric equipment. conical flask

(b) **Reagents:** Buffer Solutions of known pH values of 4.0, 7.0 and 10.0

Procedure

1. Place 10g of the test sample in a dry conical flask and add 100 ml of cool, recently boiled distilled water.

2. Make a uniform suspension, free from lumps. Allow suspension to stand at 25°C for 30 minutes, agitating continuously or intermittently in such a manner as to keep the starch particles in suspension. Let it stand for 10 more minutes.

3. Decant the supernatant liquid into the electrode vessel and immediately determine pH using a potentiometer and electrodes which have been calibrated against known buffer solutions.

Results and Inference: Record the pH value with temperature.

Precautions

• The temperature of test sample solution during the time of analysis should reported along with pH value.

• Boiled and cooled distilled water should be used for the dilution/ dispersion of sample.

• Oil and grease may interfere by coating the pH electrode and causing a sluggish response. These coatings can usually be removed by gentle wiping or detergent washing, followed by distilled water rinsing. An additional treatment with hydrochloric acid (1 per cent) may be necessary to remove any remaining film.

Ref.: Food Analysis by Neilson,4th edition, IGNOU notes.

Experiment-21

Aim: Determination of pH of the Aqueous Extract of the Sample.

Introduction: pH of the dough has marked effect on the fermentation in bread, crumb colour etc. It can be determined either by the pH meter with glass electrode or by a suitable pH comparator provided with standard colour disk. However, for more definite result, it is advisable to use pH meter.

Apparatus : pH-strips (wide range-narrow range).

Material preparation: If the sample is in liquid form, use directly for the pH estimation, but for the other samples, grind about 10 gm of material to a fine paste in a glass mortar. Add 100 ml of water and mix thoroughly. Allow to stand the mixture for about 15 minutes. Filter the mixture and collect the filtrate in another beaker. That is used as on aqueous extract.

Procedure

1. Take a small piece of wide range pH-strip and keep a small drop of aqueous extract.
2. Allow to dry for 2 to 3 minutes, so the colour become perfect.
3. Compare the colour with the standard colour chart given on the wide range strip.
4. Match the exact colour, which is the narrow range pH.
5. Match the exact colour, which is the rough pH of the aqueous extract.
6. Now select the narrow range pH in such a way, that the wide-range pH obtained is remain approximately at the middle range of narrow range pH-strip.
7. Now again make a piece of narrow range strip and keep a drop of aqueous solution on that.
8. Allow drying for 2 to 3 minutes and compare the colour developed with the standard colour chart given on the strip.
9. Find out the perfect match colour, the number written on that colour indicates the pH of the sample.

Observations

Sample No.	Wide range pH	Narrow range pH	Conclusion (acidic/basic/neutral)
1			
2			
3			
4			
5			

Experiment-22

Aim: Determination of Purity of Sodium Chloride (NaCl) Salt.

Reagents

1. 0.1 N $AgNO_3$ (Silver Nitrate).
2. 5 per cent of Potassium Chromate Indicator.

Procedure

1. Take 0.1 gm to 0.2 gm of salt sample in a stopped conical flask.
2. Add 50 ml of distilled water.
3. Add 1 ml of potassium chromate indicator.
4. Titrate against 0.1 N $AgNO_3$ (Silver Nitrate), till yellowish orange colour.
6. Note down the titre value in ml.

Calculations

Sodium Chloride (as NaCl), per cent by weight =

$$\frac{5.845 \times \text{Volume} \times \text{Normality}}{\text{Weight of Sample}}$$

Experiment-23

Aim: Determination of refractive index of fat and oil sample.

Principle: Refractive index (RI) oil or fat is a mean for identification of nature of a particular oil due to the difference of saturation, conjugation, presence of hydroxyl substituted and chain length of fatty acids. Measurement of RI is useful for the measurement of the progress of hydrogenation of oils and fats. RI is very specific for a particular oil or fat. The RI is measured under different temperature conditions such as 20°C for oils, 40° for solid fats which are fully molten at that temperature, 60°C for hydrogenated fats and 80°C for waxes. The refractive index (RI) of an oil, syrup, or other liquid is a dimensionless constant that can be used to describe the nature of the food. When a beam of light is passed from one medium to another and the density of the two differs, then the beam of light is bent or refracted. Bending of the light beam is a function of the media and the sines of the angles of incidence and refraction at any given temperature and pressure and is thus a constant (Fig.23.1).

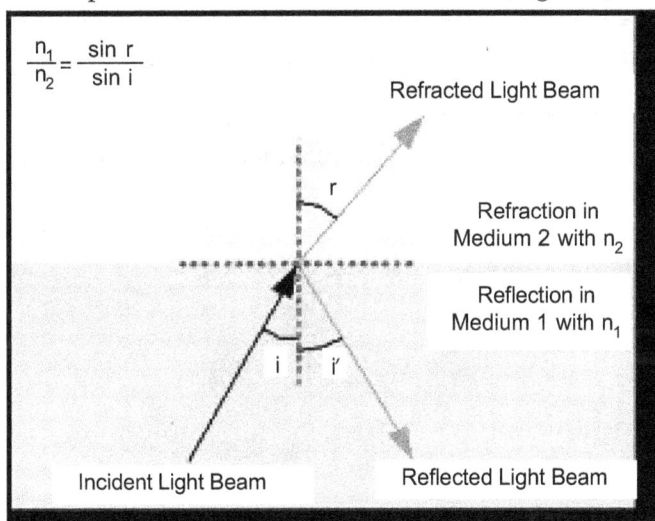

$$\frac{n_1}{n_2} = \frac{\sin r}{\sin i}$$

Refracted Light Beam

r

Refraction in Medium 2 with n_2

Reflection in Medium 1 with n_1

i   i'

Incident Light Beam   Reflected Light Beam

Fig. 23.1

Fig. 23.1 showing reflection and refraction concept of refractometry. The RI (η) is a ratio of the sines of the angles:

$$\eta = \frac{\text{sine incident ray angle}}{\text{sine refracted ray angle}}.$$

Apparatus

1. **Refractometer:** Abbe or Butyro refractometer. The temperature of the refractometer should be controlled to within ± 0.1°C and for this

purpose it should be provided with a thermostatically controlled water bath and a motor driven pump to circulate water through the instrument. The instrument should be standardized, following the manufacturer's instructions, with a liquid of known purity and refractive index or with a glass prism of known refractive index. Distilled water, which has a refractive index of 1.3330 at 20.0°C, is a satisfactory liquid for standardization.

2. **Light Source:** If the refractometer is equipped with a compensator, a tungsten lamp or a daylight bulb may be used. Otherwise, a monochromatic light, such as an electric sodium vapour lamp, should be used.

Procedure

1. Melt the sample, if it is not already liquid, and filter through a filter paper to remove any impurities and the last traces of moisture. Make sure that the sample is completely dry.

2. Adjust the temperature of the refractometer to 40.0± 0.1°C or to any other desired temperature.

3. Ensure that the prisms are clean and completely dry, and then place a few drops of the sample on the lower prism. Close the prisms, tighten firmly with the screw-head, and allow to stand for one or two minutes. Adjust the instrument and light to obtain the most distinct reading possible, and determine the refractive index.

Temperature Corrections: Unless the correction factors are specified in the detailed specification, approximate corrections shall be made using the following equation:

$$R = R' + K (T' - T)$$

where,

R = reading of the refractometer reduced to the specified temperature, $T°C$;

R' = reading at $T'°C$;

K = constant, 0.000365 for fats, and 0.000385 for oils (if Abbe refractometer is used), or =0.55 for fats and 0.58 for oils (if Butyro refractometer is used);

T' = the temperature at which the reading R' is taken; and

T = specified temperature (generally 40.0°C).

Conversation of Butyro Refractometer Readings to Refractive Indices: When Butyro refractometer is used, its readings shall be converted into refractive indices using conversion table.

Results and Inference: The mean of the results of two determinations should be taken and the temperature maintained should be reported along with the refractive index. The difference between the results of two determinations carried out simultaneously or in rapid succession shall not exceed 0.1 per cent by mass. The refractive index of commonly used edible oils is given in table below, and result of sample oil can be inferred with the standard.

Oil Sample	RI at 40°C
Coconut oil	1.448-1.449
Cottonseed oil	1.463-1.466
Ground nut oil	1.462-1.464
Mustard Oil	1.465-1.467
Sesame Oil	1.464-1.467
Safflower Oil	1.467-1.469
Sunflower Oil	1.464-1.480
Soybean Oil	1.465-1.471
Rice bran Oil	1.460-1.470
Palm Oil	1.449-1.458

Precautions

1. Before the measurement, the prism should be properly cleaned with acetone or with suitable solvent.
2. Sample should be free from moisture and any other impurities melted at the same temperature at which reading is to be taken.
3. Temperature of prism should be accurately maintained.

Ref: Food Analysis by Neilson, 4th edition, SP: 18 (Part XIII) – 1984, IGNOU notes.

Experiment-24

Aim: Determination of acid value in the fat or oil sample.

Principle: The acid value is determined by directly titrating the material in an alcoholic medium with aqueous sodium or potassium hydroxide solution. Acid Value is the number of mg of KOH required to neutralize the free fatty acids present in 1g of the oil or fat. Free fatty acid is calculated as oleic, lauric, ricinoeic or palmitic acids. Acid value is a measure of the hydrolytic rancidity present in the sample.

Requirements

(a) **Apparatus:** Erlenmeyer Flask of 250 ml capacity.

(b) **Reagents**

1. Ethyl Alcohol (95 Percent v/v) or rectified Spirit neutral to phenolphthalein indicator.
2. Phenolphthalein Indicator Solution – Dissolve 1 g of phenolphthalein in 100 ml of ethyl alcohol.
3. Standard Aqueous Potassium Hydroxide or Sodium Hydroxide Solutions - 0.1 N or 0.5 N.

Note: When testing oils or fats which give dark coloured soap solution, the observation of the end point of the titration may be facilitated either (a) by using thymolphthalein or alkali blue 6B in place of phenolphthalein, or (b) by adding 1 ml of a 0.1 percent (w/v) solution of methylene blue in water to each 100 ml of phenolphthalein indicator solution before titration.

Procedure

1. Mix the oil or melted fat thoroughly before weighing.
2. Weigh accurately a suitable quantity of the cooled oil or fat in a 200 ml conical flask. The weight of the oil or fat taken for the test and the strength of the alkali used for the titration shall be such that the volume of alkali required for the titration does not exceed 10 ml.
3. Add 50 to 100 ml of freshly neutralized hot ethyl alcohol, and about 1 ml of phenolphthalein indicator solution. Boil the mixture for about five minutes and titrate while as hot as possible with standard aqueous alkali solution shaking vigorously during titration.

Calculation

Acid value = $56.1 \times V \times N/W$.

where,

> V = Volume in ml of standard potassium hydroxide or sodium hydroxide solution used,
>
> N = Normality of standard potassium hydroxide or sodium hydroxide solution, and
>
> W = Weight in g of the material taken for the test.

Results and Inference: Report the acid value of sample obtained. The difference between the results of two determinations carried out simultaneously or in rapid succession should not exceed 0.1. Infere the result based on the standard value as mentioned in the table below:

OIL sample	Acid Value
Coconut oil	0.5
Cottonseed oil	0.3
Ground nut oil	0.5
Mustard Oil	0.5
Sesame Oil	0.5
Safflower Oil	2.0
Sunflower Oil	0.5
Soybean Oil	0.5
Rice bran Oil	0.5
Palm Oil	0.5

Table showing standard acid value of oil.

Precautions

1. The formation of two layers should be avoided by vigorous shaking so that the free acids do not get transferred into the ethanolic layer.
   2.The freshly neutralized alcohol must able be hot at the time of addition.
3. The weight of the oil or fat taken for acidity determination and the strength of NaOH should be such that the volume of alkali used does not exceed 10ml.

Ref: Food Analysis by Neilson, 4th edition, SP : 18 (Part XIII) – 1984, IGNOU notes.

Experiment-25

Aim: Determination of free fatty acids in the fat oil sample.

.

Principle: Measures of fat acidity normally react the amount of fatty acids hydrolyzed from triacylglycerols. As per following equation:

Triacylglycerol + $3H_2O$ ⟶ Glycerol + Fatty Acids

FFA is the percentage by weight of a specied fatty acid (e.g., percent oleic acid). The acidity is frequently expressed as the percentage of free fatty acids present in the sample. The percentage of free fatty acids in most of the oils and fats is calculated on the basis of oleic acid; although in coconut oil and palm kernel oil it is often calculated in terms of lauric acid, in castor oil in terms of ricinoleic acid, and in palm oil in terms of palmitic acid.

Requirements

(a) **Apparatus** : Erlenmeyer Flask of 250 ml capacity.

(b) **Reagents**

1. Ethyl Alcohol (95 Percent v/v) or rectified Spirit neutral to phenolphthalein indicator.
2. Phenolphthalein Indicator Solution – Dissolve 1 g of phenolphthalein in 100 ml of ethyl alcohol.
3. Standard Aqueous Potassium Hydroxide or Sodium Hydroxide Solutions - 0.1 N or 0.5 N.

Note: When testing oils or fats which give dark coloured soap solution, the observation of the end point of the titration may be facilitated either (a) by using thymolphthalein or alkali blue 6B in place of phenolphthalein, or (b) by adding 1 ml of a 0.1 per cent (w/v) solution of methylene blue in water to each 100 ml of phenolphthalein indicator solution before titration.

Procedure

1. Mix the oil or melted fat thoroughly before weighing.
2. Weigh accurately a suitable quantity of the cooled oil or fat in a 200ml conical flask. The weight of the oil or fat taken for the test and the strength of the alkali used for the titration shall be such that the volume of alkali required for the titration does not exceed 10 ml.
3. Add 50 to 100 ml of freshly neutralized hot ethyl alcohol, and about 1 ml of phenolphthalein indicator solution.

4. Boil the mixture for about five minutes and titrate while as hot as possible with standard aqueous alkali solution shaking vigorously during titration.

Calculations

FFA and acid value may be converted from one to the other using a conversion factor equation. Acid value conversion factors for lauric and palmitic are 2.81 and 2.19 respectively

per cent FFA (as oleic acid) ×1.99= acid value

The calculations in terms of different fatty acids are as follows:

(a) Free fatty acids, in terms of oleic acid, percent by weight

$$= 26.2 \times V \times N / W$$

(b) Free fatty acids, in terms of lauric acid, percent by weight

$$= 20.0 \times V \times N / W$$

(c) Free fatty acids, in terms of ricinoleic acid, percent by weight

$$= 29.8 \times V \times N / W$$

(d) Free fatty acids, in terms of palmitic acid, percent by weight

$$= 25.6 \times V \times N / W$$

where,

V = volume in ml of standard potassium hydroxide solution used,

N = normality of standard potassium hydroxide solution, and

W = weight in g of the material taken for the test.

Results and Inference: Report the value of FFA obtained. The difference between the results of two determinations carried out simultaneously or in rapid succession should not exceed 0.1. Infere the result based on the standard value.

Precautions

1. The formation of two layers should be avoided by vigorous shaking so that the free acids do not get transferred into the ethanolic layer.
2. The freshly neutralized alcohol must able be hot at the time of addition.
3. The weight of the oil or fat taken for acidity determination and the strength of NaOH should be such that the volume of alkali used does not exceed 10ml.

Ref.: Food Analysis by Neilson, 4th edition, SP : 18 (Part XIII) – 1984, IGNOU Notes.

www.ingramcontent.com/pod-product-compliance
Lightning Source LLC
Chambersburg PA
CBHW050229270326
41914CB00003BA/626